PROGRESS IN SUPERCONDUCTIVITY RESEARCH

PROGRESS IN SUPERCONDUCTIVITY RESEARCH

OLIVER A. CHANG
EDITOR

Nova Science Publishers, Inc.
New York

For permission to use material from this book please contact us:
Telephone 631-231-7269; Fax 631-231-8175
Web Site: http://www.novapublishers.com

LIBRARY OF CONGRESS CATALOGING-IN-PUBLICATION DATA

Progress in superconductivity research / Oliver A. Chang, editor.
p. cm.
Includes index.
ISBN-13: 978-1-60021-694-7 (hardcover)
ISBN-10: 1-60021-694-3 (hardcover)
1. Superconductivity--Research. I. Chang, Oliver A.
QC611.96.P76 2006
621.3'5--dc22 2007016764

Published by Nova Science Publishers, Inc. ✢ New York

CONTENTS

PREFACE

This new book presents state-of-the art research on superconductivity which is the ability of certain materials to conduct electrical current with no resistance and extremely low losses. High temperature superconductors, such as La2-xSrxCuOx (Tc=40K) and YBa2Cu3O7-x (Tc=90K), were discovered in 1987 and have been actively studied since. In spite of an intense, world-wide, research effort during this time, a complete understanding of the copper oxide (cuprate) materials is still lacking. Many fundamental questions are unanswered, particularly the mechanism by which high-Tc superconductivity occurs. More broadly, the cuprates are in a class of solids with strong electron-electron interactions. An understanding of such "strongly correlated" solids is perhaps the major unsolved problem of condensed matter physics with over ten thousand researchers working on this topic.

Gauge effects on the fluctuation properties of the normal-to-superconducting phase transition in bulk and thin film superconductors are reviewed in Chapter 1. Similar problems in the description of other natural systems (liquid crystals, quantum field theory, early universe) are also discussed. The relatively strong gauge effects on the fluctuations of the ordering field at low spatial dimensionality D and, in particular, in thin (quasi-2D) films are considered in details. A special attention is paid to the fluctuations of the gauge field. It is shown that the mechanism, in which these gauge fluctuations affect the phase transition order and other phase transition properties varies with the variation of spatial dimensionality D. The problem for the experimental confirmation of theoretical predictions about the order of phase transitions in gauge systems is discussed. Related topics: gauge effects on the critical behavior of unconventional superconductors, disorder, quantum fluctuations in a close vicinity of ultra-low phase transition temperatures , are also briefly discussed.

In Chapter 2, the Excitonic Enhancement Mechanism [1] is proposed as the theory for superconductivity in some novel hole carrier systems, such as M_3C_{60} [2], cuprate high temperature superconductors [3] as well as the most recently discovered Boron doped Silicon [4]. It is suggested that because excitons are like hydrogen atoms and possess excited energy levels, these Cooper pair exchange particles can be excited by suitable infrared light [5], resulting in large enhancement of the superconductivity transition temperature [5]. Experimental data on the EEM Tc enhancement was shown for bulk and thin film YBCO. With the optimum YBCO bulk system (inside an open liquid Nitrogen cryogenic chamber), the authors show the enhancement by observing the diamagnetism before and after IR light was applied. The transition temperature was shown to increase ~80°K, from $93^{o}K$ *to* $\geq 175^{o}K$ exactly as predicted by the EEM theory [5]. The results of these original measurements were verified on an oxygen deficient YBCO thin film by measuring the transition temperature using AC voltage versus temperature measurement (inside a closed Helium cryogenic chamber) carefully obtained by using a 100Khz DSP lock-in-amplifier, before and after IR light was applied. The transition temperature was shown to increase by ~17°K, from 69.5°K to 86.5°K[5]. The increase in Tc, although small, is shown to be exactly as the value calculated by the same EEM theory [5]. It is therefore without a doubt that these results confirm EEM as the theory for cuprates and its application to high Tc superconductors.

In Chapter 3, the authors derive a new boundary-value problem for the ferromagnetic metal/superconductor (FM/S) nanostructures, where superconductivity is a superposition of the BCS pairing with zero total momentum in the S layers and the Fulde-Ferrell-Larkin-Ovchinnikov (FFLO) pairing with nonzero 3D coherent momentum **k** in the FM layers. It is shown that processes of mutual transformation between the BCS and FFLO pairs at the S/FM boundary go via the unrealized *virtual states* (FFLO and BCS, correspondingly). These processes occur as the *Umklapp processes* during which excess coherent pair momentum **k** becomes confined in the FM layer. This fact is reflected in the significantly new boundary conditions for the Eilenberger equations. On this base they originally investigate the interplay between the BCS and 3D FFLO states in the pure FM/S bilayers. In the FM/S/FM trilayers two novel *π phase superconducting states with electron-electron repulsion* in the FM layers are predicted. The 3D modulated FFLO states are possible in such trilayers only in presence of external magnetic field at suitable parameters of the FM and S layers. In the FM/S superlattices there are also two π phase magnetic states (0π and ππ) with compensation of the exchange field paramagnetic effect. This fact allows us to explain a surprisingly high $T_c \sim 5$K in the short period Gd/La superlattice and to predict the sign and value of the electron-electron interaction in the Gd metal.

In Chapter 4, heavily boron-doped polycrystalline diamond thick films were synthesized via the electron-assisted hot filament chemical vapor deposition method. The surface morphology and thickness for a freestanding polycrystalline diamond film were evaluated by scanning electron microscope (SEM), and the structure and phase purity of as-grown heavily boron-doped diamond film were characterized by X-ray diffraction (XRD) and micro-Raman spectroscopy. The film resistivity and the carrier concentration as well as the relation of the resistivity to carrier concentration were obtained at room temperature by the van der Pauw method and the Hall-effect method, respectively. Furthermore, the properties of electrical transport and magnetization were measured between room temperature and 1.9K, and the results indicated the superconducting transition temperatures were 10 K for onset and 8.3 K for zero resistance as well as the upper critical field was 2.7 T. Subsequently, the authors discussed the possible factors of inducing such a high Tc value for as-grown heavily boron-doped polycrystalline diamond film. In particular, the contrastive studies of optical response properties for present boron-doped diamond film and the undoped diamond film substantiated that there was the strong electron-phonon coupling in the heavily boron-doped diamond system.

The high-T_c cuprates have been a focus of intense study since its discovery in 1986. The 20th anniversary of the discovery of this material is marked by a partial understanding of its properties. They are far from being fully understood. The cuprates not only demonstrate a very high T_c, but, also exhibit various anamolies in their superconducting and the normal state properties. The pairing symmetry of the Cooper pairs was a topic of intense debate for more than a decade. A new class of phase sensitive tests as well as various other techniques for energy-gap measurements have revealed a predominantly d-wave pairing symmetry in the cuprates unlike the *s*-wave states in conventional superconductors. At present one of the most peculiar and controversial topics in these materials is the phenomenon of pseudogap in the normal state of the cuprates. A depletion of the low-energy density of states is observed in the underdoped regime of these materials by various experimental measurements like the optical conductivity, ARPES, c-axis tunneling, specific heat etc. The pseudogap has also been observed to have the same k-dependence as that of the superconducting gap. A number of theoretical scenarios have been proposed to explain the origin of this phase. It was initially believed to be a crossover phenomenon where the pseudogap had been ascribed as a natural consequence of local, dynamic pairing correlations in the normal state of low-density small pair size superconductors. Apart from the other suggestions like the phase fluctuations of the superconducting order parameter in the pseudogap regime, recently, the *d*-densiy wave (DDW) scenario has been proposed as a strong candidate to explain the pseudogap. In fact, various theoretical works on DDW have been successful in explaining some of the salient features of this phase. A number of experiments are also consistent with this proposal. In Chapter 5 the authors will present a brief overview of the phase diagram of the cuprates and the present experimental situation followed by a discussion on the present status of the DDW order as a candidate for the pseudogap. They will present a comparison between the DDWorder and the other scenarios. The authors will also explain, in short, the conventional density-wave order and its difference with the unconventional or DDW order. The authors will discuss the theoretical findings of the DDW order in the limelight of the available experimental predictions. Some proposals of experimental measurements which might play an important role to strengthen the candidature of DDW order will also be considered.

Superconductivity in a system of strongly correlated electrons treated as a singular Fermi liquid (SFL) is studied. In the system the scattering amplitude for particles with the same momenta is divergent. The qualitatively new ground state and other eigenstates for the superconducting phase are found. Thermodynamic functions are derived within the Bogolubov method. Analytical formulae for the free energy as well as the gap equation and the equation for the chemical potential are presented in Chapter 6. The gap equation and the equation for the chemical potential are found to be mutually correlated and have to be simultaneously solved in a self-consistent manner. The numerical solution for the case of the *s*-wave paired SFL is discussed. It is found that the energy gap and the chemical potential are doubly-valued in the subcritical region, however only the upper branches are thermodynamically stable.

According to the common property of temperature-doping phase diagrams for high- T_c superconductors in the underdoped regime, two characteristic temperatures can be identified in the SFL model. The higher one is the temperature T^* at which phase-incoherent Cooper pairs appear and the pseudogap emerges. The other one is T_c, lower than T^*, at which the Cooper pairs become phase-coherent and form a superconducting condensate. The obtained results evidence that the SFL system displays some features of the superconducting state also in the supercritical temperature range, $T_c < T < T^*$: The non-zero value of the energy gap points at a possibility of Cooper pairs formation, although in terms of the free energy the preferred phase is the normal one. Thus, the supercritical region with a non-zero energy gap can be related to the emergence of the pseudogap. The phase transition to the superconducting state is found to be of the first order with a pronounced discontinuity in the entropy. The hysteresis phenomenon in the supercritical temperature range is considered in detail.

In order to study the response of the system to a weak magnetic field acting on both spin and charge degrees of freedom, the SFL model is extended by including the Pauli term and the superflow. In the extended model the eigenstates for the superconducting phase are found. The spin susceptibility and the magnetic field penetration depth are discussed. It is shown that the normal-phase spin susceptibility has a Curie-law character. Critical parameters such as the critical magnetic field and the critical current are studied employing the variational expansion of the free energy.

EXPERT COMMENTARIES

In: Progress in Superconductivity Research
Editor: Oliver A. Chang, pp. 3-5
ISBN: 1-60021-694-3

MUTUAL ACCOMMODATION OF SUPERCONDUCTIVITY AND MAGNETISM IN LAYERED FERROMAGNET-SUPERCONDUCTOR NANOSTRUCTURES

***Yuri A. Izyumov*[1], *Mansur G. Khusainov* [2-4] *and Yurii N. Proshin* [2,3,*]**

[1)] Institute of Metal Physics, Ural Division of RAS, Ekaterinbourg 620219, Russia
[2)] Kazan State University, Kazan 420008, Russia
[3)] Max-Planck-Institute for the Physics of Complex Systems, Dresden 01871, Germany
[4)] Kazan State Tupolev Technical University, Chistopol' 422981, Russia

Keywords: superconductivity, ferromagnetism, proximity effect, mutual accommodation, boundary value problem, Umklapp processes

PACS: 74.80.Dm; 74.50.+r; 74.62.-c

The goal of this commentary is briefly to sum up the fifteen-year activity of theoretical investigations of the mutual accommodation of the superconducting and magnetic states in the layered FM/S systems where the superconducting (S) and ferromagnetic metal (FM) regions are spatially separated but coupled with one another through the interfaces between the layers.

In the time passed since the publication of the pioneering works by Buzdin, Radović et al. [1,2], in which the fundamentals of the theory of the layered FM/S nanostructures were developed, numerous theoretical and experimental investigations of these heterogeneous systems have been carried out (see recent reviews [3-6] and references therein). For

[*] E-mail address: Yurii.Proshin@ksu.ru

conditions corresponding to the most experiments (dirty superconductor, when the coherence length is less than respective layer thickness), it was shown that such systems are described by a boundary-value problem for a pair amplitude (the Usadel function). In such a heterogeneous system a superconductivity proves to be a superposition of the superconductivities of the Bardeen-Cooper-Schrieffer (BCS) type in the S layers and of the Fulde-Ferrel-Larkin-Ovchinnikov (FFLO) [7,8] type in the FM layers. The last type of inhomogeneous pairing is characterized by nonzero momentum of the superconducting pair. The most important success of the theory is the explanation of the nonmonotonic dependence of the critical temperature T_c on the thickness of the ferromagnetic layer d_f [1-4] and the prediction of the reentrant superconductivity [9], that was observed not long ago [10]. Oscillations of T_c are caused by conditions at the boundary: the flux of Cooper pairs is proportional to the jump in the pair amplitude at the ferromagnet-superconductor interface. The solution of the boundary-value problem makes it possible to calculate T_c as a function of d_f. Depending on the microscopic parameters that characterize the S layer, the FM layer, and the boundary, varied behavior of the FM/S systems can be expected and is observed in numerous experiments (see references in reviews [3-6]).

Another problem concerning the behavior of the FM/S systems (of superlattices, in particular) is the problem of the mutual orientations of the magnetizations of the FM layers as an effect of the inverse influence of superconductivity on magnetism (see reviews [3-4] and references therein). Analysis of the boundary-value problem makes it possible to find conditions under which a ferromagnetic or antiferromagnetic arrangement of the magnetizations of the FM layers is realized (the antiferromagnetic orientation proves to be more favorable). Thus, the theory of the few-layered FM/S systems and the FM/S superlattices based on the boundary-value problem for the Usadel function is capable of explaining the experimentally observed behavior of these systems. Of particular interest is the case of four-layered system FM/S/FM'/S' in which the theory predicts decoupled superconductivity in layers S and S' [11]. A rising hierarchy of critical temperatures leads to the larger set of states (up to seven!) in comparison with well-known FM/S/FM trilayer.

So far, we have spoken of the thermodynamic properties of FM/S systems. Now, we summarize the investigations of the main features of their transport properties. They manifest themselves most vividly in three-layer junctions. Thus, an S/FM/S junction can most conveniently be used to study π-phase superconductivity by the observation of oscillations of the Josephson current across the junction (see reviews [5-6] and references therein). Vanishing of the critical current at a certain thickness of the ferromagnetic layer (or at a certain temperature if d_f is fixed) with its subsequent increase is persuasive evidence of the existence of π-phase superconductivity in such systems.

Another class of phenomena arises in the FM/S/FM junctions depending on the mutual orientations of magnetizations of the FM layers [3-6]. For the antiferromagnetic orientation, a nonequilibrium spin polarization due to the injection of magnetized electrons from the ferromagnetic layer is accumulated in the superconducting layer. As a result, the superconducting characteristics of the entire junction, e.g., the superconducting gap or the tunneling magnetoresistance, depend quite specifically on the applied voltage.

Of large interest is the investigation of the proximity effect in the *clean* FM/S nanostructures, which could be used as ideal model systems for direct observation of a realization of the unconventional FFLO state and its competition with the conventional BCS type of pairing. But the theory of pure FM/S systems remains to be constructed. The equations for the Eilenberger function, which represent the proximity effect in the clean FM/S systems, are derived in this issue [12]. This theory allows us to explain the surprising absence of the superconductivity suppression in the short period Gd/La superlattices [13], where the thickness of ferromagnetic gadolinium was nearly triply as large as the thickness of superconducting lanthanum. In addition, on the base of such investigations of the FM/S bilayers and FM/S/FM trilayers, one can propose the *method of the proximity effect spectroscopy* [12], which makes possible the detection of unknown electronic parameters of the FM metal (the exchange field and the interelectronic interaction) if well-known BCS superconductor S is used as a probe. The extension of our results on the high temperature superconductors (HTS) and the study of the S/HTS and FM/HTS systems can provide a good method for a probe of the electronic correlations type and the order parameter symmetry in the HTS material.

Thus, owing to intense theoretical and experimental investigations, a significant understanding of the effects of the mutual influence of superconductivity and magnetism in the heterogeneous FM/S systems has been achieved. There are also numerous suggestions for engineering applications of some of the above considered phenomena in the *superconducting spintronics*, in particular, for the fabrication of electric current switches, in which the switching is effected using a weak magnetic field, or for the production of devices based on quantum bits of information. This stimulates the further study of layered FM/S systems.

There is no doubt that the problems of the coexistence of superconductivity and ferromagnetism will in the near future be at the peak of interest of theoreticians, experimentalists, and research engineers.

This work was supported in part by the Russian Foundation for Basic Research (Grants 05-02-16369, 08-02-01376).

References

[1] Radović Z. *et al. Phys. Rev. B* 1991, 44, 759-764.

[2] Buzdin A.I.; Vujićić B.; Kupriyanov M.Yu. *Sov. Phys. JETP* 1992, 74, 124-133.

[3] Izyumov, Yu.A.; Proshin, Yu.N.; Khusainov, M.G. *Physics-Uspekhi* 2002, 45, 109-148.

[4] Buzdin, A.I. *Rev. Mod. Phys.* 2005, 77, 935-976.

[5] Golubov, A.A.; Kupriyanov, M.Yu.; Il'ichev, E. *Rev. Mod. Phys.* 2004 76, 411-469.

[6] Ryazanov, V.V. *et al. J. Low Temp. Phys.* 2004, 136, 385-400.

[7] Fulde, P.; Ferrell, R.A. *Phys. Rev.* 1964, A135, 550-563.

[8] Larkin, A.I.; Ovchinnikov, Yu.N. *Sov. Phys. JETP* 1965, 20, 762-771.

[9] Khusainov, M.G.; Proshin, Yu.N. *Phys. Rev. B* 1997, 56, 14283-14286.

[10] Garifullin, I.A. *et al. Phys. Rev. B* 2002, 66, 020505.

[11] Proshin, Yu.N.; Zimin, A; Fazleev, N.G.; Khusainov, M.G. *Phys. Rev B* 2006, 73, 184514

[12] Khusainov M.G.; Khusainov M.M.; Proshin Yu.N. *This issue*, Chapter 3

[13] Goff, J.P. *et al. J. Magn. Magn. Mater.* 2002, 240, 592-594.

In: Progress in Superconductivity Research
Editor: Oliver A. Chang, pp. 7-10
ISBN 1-60021-694-3

On Peculiarities of Multiband Superconductivity with the Interband Pairing

Nikolai Kristoffel*
Institute of Physics, University of Tartu, Riia 142, 51014 Tartu, Estonia
Institute of Theoretical Physics, University of Tartu, Tähe 4, 51010 Tartu, Estonia

The multiband superconductivity incorporating the interband pairing channel is known for a long time [1,2]. This framework contains numerous peculiarities, novel events and manifestations as compared with the conventional Bardeen-Cooper-Schriffer scheme. The understanding of superconductivity extends to materials with complex inner structure and varied multicomponent electronic spectra. The conditions for the appearance of superconductivity become extended by functioning of nonconvenient interaction channels. The strength of the pairing can be expected to be amplified.

The interest to multiband superconductivity has grown particularly in connection with the difficulties in explaining the pairing mechanism in cuprate superconductors. Various connected aspects and diverse model realizations have been investigated, e.g. [3-9]. The first widely accepted successful application of the two-band model to a real system has been done for the two-band superconductor MgB_2 [10-13].

At the present time the nomenclature of the classes of materials under interest and proved for the superconductivity is wide. Cuprate perovskites with all the relative metal oxides, the MgB_2 family, heavy fermion and A15 type compounds, charge-transfer organic materials, graphite intercalation compounds, fullerides, carbon nanotubes, etc. can be mentioned. One deals here with complex materials for which a multiband electron spectrum near the Fermi energy has been found or expected. Analogous quantum resonance conditions can be reached also by special impacts, i.e. doping treatments, external pressure, or nanoarchitectural constructions. Accordingly the applications of multiband superconductivity schemes to more or less exotic superconductors is recently becoming popular. However, the features which can be simply interpreted using the multiband superconductivity approaches seem to be not well and largely known. For this reason some peculiarities

*E-mail address: kolja@fi.tartu.ee

of multiband superconductivity with the interband pairing channel will be mentioned in this comment.

First of all, the interband pairing mechanism (off-diagonal exchange-like pairing) permits to reach in a natural way high superconductivity transition temperatures. The active volume of the momentum space can be here considerable at the profitable disposition of the chemical potential crossing multiple overlapping bands. This mechanism induces superconductivity also by repulsive interband coupling [14] and Coulombic forces besides (interband) electron-phonon interactions can participate in the pairing.

The multicomponent order parameter describes interfering superconducting condensates with own supercarrier densities and phases. Supercarrier densities and the corresponding superconducting gaps of the band components are considerably determined by the effectiveness of the interband pairing. The change in the corresponding conditions (by doping, pressure, etc.) can lead to peculiar behaviour of the transition temperature. For example, unfavourable to favourable and the opposite change in interband pairing conditions can lead to a bell-like behaviour of T_c [3,8] on the phase diagram.

The symmetry of the pairing is not restricted to the s-wave in the multiband schemes. The superconducting gaps (Δ) of different bands can differ in symmetry. The pairing symmetry can change over the phase diagram with doping etc. [15]. In case of two totally symmetric order parameters and repulsive interband interaction the corresponding gaps are of different signs [16].

Both superfluids (a two-band system) have the same critical temperature in the presence of interband pairing and both superconducting gaps vanish at this T_c simultaneously. The characteristic ratios Δ_α/k_BT_c can deviate markedly from the BCS universality [7].

The low-energy excitation spectrum of a multiband superconductor can manifest superconducting gaps and pseudogaps. The latter appear naturally as minimal quasiparticle excitation energies of band components not intersected by the chemical potential (μ) [7,8]. Then the pseudogap energy is given by the expression $\sqrt{(\epsilon_\alpha - \mu)^2_{min} + \Delta^2_\alpha}$ and this gap survives in the normal state as $|\epsilon_\alpha - \mu|_{min}$. Such "extrinsic" source of the pseudogap does not manifest itself in the supercarrier density because the interband pairing channel can induce the superconductivity also for the gapped bands. On the phase diagram, when an external impact (e.g. doping) reorganizes the spectral relations so that $\epsilon_\alpha = \mu$ is reached, the pseudogap transforms into the superconducting gap.

In multiband superconductors a novel source of the phase diagram critical points is opened corresponding to the starting overlap of the band components [7,8].

One must be careful in discussing the multigap spectra. Note that the smaller gap (as compared with the other superconducting gap, or the pseudogap) can manifest indefinitely as an additional density in the V-like joint spectral hole. The different behaviour of gaps in the momentum space can also introduce difficulties for interpretations.

The strength of the pairing and the phase coherence are expected to develop and vanish simultaneously due to the interband interference [17]. In two-component systems the critical superconductivity order fluctuation mode (Goldstein type) nonzero equilibrium value at $T < T_c$ determines both the superconducting gaps. Its fluctuations are exponentially quenched in space. As a novel event an additional noncritical fluctuation mode driven by the phase difference of two superfluids (the Leggett mode [18]) appears. A spatially periodic coherence wave (with an imaginary correlation length) corresponds to its fluctuations.

Such a system possesses also two relaxation channels with critical and noncritical characteristics times [19].

The paired carrier number density is essentially determined by the paired carrier density $n_s(T)$. In multiband systems contrary to the BCS one-band case $n_s(0) \neq 1$, i.e. not all the carriers become paired at $T = 0$. This circumstance plays an essential role in analysing the penetration depth [17], because the mechanism determining n_s can be influenced e.g. by isotope substitution. Note that the interband electron-phonon interaction constant $V \sim M^{-1}$, where M is the effective mass of the active vibration (in the one-band case the intraband $V = const$). Accordingly one predicts the isotope effect on the paired carrier density in multiband superconductors [20]. This must be taken into account when analysing the isotope effect on superconducting stiffness etc. The approximate expressions for T_c in the presence of interband pairing reflect the opposite trend of T_c and its isotope effect exponent on the phase diagram [3]. A moderate contribution of the electron-phonon interaction into the whole interband coupling, being not decisive in the pairing, can nevertheless cause "normal valued" isotope exponents.

The multiplicity of the order parameters and the interaction of them influences the temperature dependences of the critical magnetic fields, e.g. the field H_{c2} shows a dependence with positive curvature. The presence of two superconducting gaps manifests in the anomalous temperature dependence of the heat capacity in the superconducting state. Such a nomenclature can be continued and has been especially developed in connection with the investigations of the MgB_2 and borocarbide family properties [21,22].

The multiband approaches are expected to be proved for the description and interpretation of superconducting order in various materials of current interest. Intentional search for novel effective superconductors can trace this way. The internal inhomogeneity of complex compounds seems to play a remarkable role at this. A topic designated as "superconductors in multiband scenarios" has already started.

Random references for the introduction into the field have been chosen.

The author is indebted to Dr. T.Örd for discussion.

This work has been supported by the Estonian Science Foundation Grant No 6540.

References

[1] Suhl, H., Matthias, B. and Walker, L., *Phys. Rev. Lett.* **3**, 552 (1959).

[2] Moskalenko, A., *Fiz. Met. Metalloved.* **8**, 503 (1959).

[3] Kristoffel, N., Konsin, P. and Örd, T., *Rivista Nuovo Cim.* **17**, 1 (1994).

[4] Bianconi, A., *Solid State Commun.* **89**, 933 (1994).

[5] Perali, A. et al., *Phys. Rev. B* **62**, R9295 (2000).

[6] Bussmann-Holder, A., Micnas, R. and Bishop, A.R., *Eur. Phys. J. B* **37**, 345 (2004).

[7] Kristoffel, N. and Rubin, P., *Solid State Commun.* **122**, 265 (2002).

[8] Kristoffel, N. and Rubin, P., *Physica C* **402**, 257 (2004); *J. Supercond.* **18**, 705 (2005).

[9] Kamimura H. et al., Theory of Copper Oxide Superconductors, Springer, Berlin, 2005.

[10] Bianconi, A., *J. Supercond.* **18**, 25 (2005). Contains an extended list of references. See also Bianconi, A., Filippi, M. In: Symmetry and Heterogeneity in High-Temperature Superconductors, Ed. A.Bianconi, Springer, 2006, p. 21.

[11] Liu, A., Mazin, Y.Y. and Kortus, Y., *Phys. Rev. Lett.* **82**, 087005 (2001).

[12] Bianconi, A., Agrestini, S. and Bussmann-Holder, A., *J. Supercond.* **17**, 205 (2004).

[13] Kristoffel, N., Örd, T. and Rägo, K., *Europhys. Lett.* **61**, 109 (2003).

[14] Kondo, J., *Progr. Theor. Phys.* **29**, 1 (1963).

[15] Konsin, P., Kristoffel, N. and Rubin, P., *phys. stat. solidi b* **208**, 145 (1998).

[16] Kristoffel, N., Örd, T. and Konsin, P., *Nuovo Cim. D* **16**, 311 (1994).

[17] Kristoffel, N., Örd, T. and Rubin, P., *Physica C* **437-438**, 168 (2006).

[18] Leggett, A.J., *Progr. Theor. Phys.* **36**, 901 (1966).

[19] Örd, T. and Kristoffel, N., *Physica C* **331**, 13 (2000).

[20] Kristoffel, N. and Rubin, P., *Phys. Lett. A* **360**, 367 (2006).

[21] Askerzade, I.N., *Uspekhi Fiz. Nauk* **176**, 1025 (2006).

[22] Mansor, N. and Carbotte, J.P., *Phys. Rev. B* **72**, 024538 (2005).

RESEARCH AND REVIEW STUDIES

In: Progress in Superconductivity Research
Editor: Oliver A. Chang, pp. 13-53

ISBN 1-60021-694-3

Chapter 1

Fluctuation and Gauge Effects on the Critical Behavior of Superconductors

Diana V. Shopova** and Dimo I. Uzunov
CP Laboratory, Institute of Solid State Physics,
Bulgarian Academy of Sciences, BG-1784 Sofia, Bulgaria, and
Max-Plank-Institut für Physik komplexer Systeme,
Nöthnitzer Str. 38, 01187 Dresden, Germany

Abstract

Gauge effects on the fluctuation properties of the normal-to-superconducting phase transition in bulk and thin film superconductors are reviewed. Similar problems in the description of other natural systems (liquid crystals, quantum field theory, early universe) are also discussed. The relatively strong gauge effects on the fluctuations of the ordering field at low spatial dimensionality D and, in particular, in thin (quasi-2D) films are considered in details. A special attention is paid to the fluctuations of the gauge field. It is shown that the mechanism, in which these gauge fluctuations affect the phase transition order and other phase transition properties varies with the variation of spatial dimensionality D. The problem for the experimental confirmation of theoretical predictions about the order of phase transitions in gauge systems is discussed. Related topics: gauge effects on the critical behavior of unconventional superconductors, disorder, quantum fluctuations in a close vicinity of ultra-low phase transition temperatures , are also briefly discussed.

Key words: thermodynamics, order parameter, specific heat, equation of state, renormalization group, fluctuation.

PACS: 74.20.De, 74.78.Db

*E-mail address: sho@issp.bas.bg. (Corresponding author)

1. Introduction

1.1. Gauge Effects and Superconductivity

A remarkable example of gauge theory in condensed matter physics is the Ginzburg-Landau (GL) functional of superconductivity [1, 2]. The latter is invariant towards both global $U(1)$ rotations of the order parameter field $\psi(\vec{x})$ and local gauge transformations (rotations) of the same field and the vector potential $\vec{A}(\vec{x})$ of magnetic induction $\vec{B}$. These two properties of gauge invariance define the global and local $U(1)$ symmetries of the GL theory. In both cases the gauge group is a one-dimensional Abelian continuous group, $U(1)$. The spontaneous breaking of these global and local symmetries below the phase transition point allows the appearance of the superconducting phases: the uniform Meissner phase and, under certain circumstances, the mixed (Abrikosov vortex [3]) phase. While the Meissner phase $[\langle\psi\rangle \neq 0,\ \langle\vec{B}\rangle = 0]$ is a mere product of the global $U(1)$ symmetry breaking, the vortex phase $[\langle\psi(\vec{x})\rangle \neq 0,\ \langle\vec{B}(\vec{x})\rangle \neq 0]$, where the equilibrium field configurations of both order parameter field $\psi(\vec{x})$ and magnetic induction $\vec{B}(\vec{x})$ are spatially nonuniform, occurs as a result of the spontaneous breaking of local $U(1)$ symmetry. The local gauge is important also for the description of magnetic field penetration in both Meissner and vortex phases. The penetration is described by an additional characteristic length, the London penetration length [1, 2], which is different from zero only when the symmetry is broken, i.e., in the ordered phases, where the equilibrium value of ψ is different from zero. The local $U(1)$ symmetry is present when the superconductor is in external magnetic field, or, when this field is equal to zero but magnetic fluctuations exist and determine the vector potential $\vec{A}(\vec{x})$ as a purely fluctuating field.

The fluctuations of relevant fields, $\psi(\vec{x})$ and $\vec{A}(\vec{x})$ in usual superconductors are small and can be neglected. Therefore, the GL functional can be investigated in the lowest order mean field (MF) approximation (alias, "tree approximation" [4, 5]). This is the usual way of treatment of the GL free energy, in particular, the study of GL equations [1, 2, 3]. Within the tree approximation, the phase transition from normal to superconducting state in a zero magnetic field is of second order [1, 2, 4].

For long time this phase transition has been considered as one of the best examples of second order phase transitions, which has an excellent description within MF. But in 1974 Halperin, Lubensky and Ma (HLM) [6] showed that the magnetic fluctuations change the order of the superconducting phase transition in a zero external magnetic field ($H_0 = |\vec{H}_0| = 0$)), i.e., the order of the phase transition from normal-to-uniform (Meissner) superconducting state at $T_{c0} = T_c(H_0 = 0)$ (see, also Ref. [7]). Since then this fluctuation change of the order of normal-to-superconducting phase transition (*HLM effect*) has been under debate. After 1974 an overwhelming amount of theoretical research on this topic has been performed but up to now there is no complete consensus about the order of this phase transition.

In this review we shall consider some aspects of this problem but we shall not be able to discuss or mention all relevant contributions. However, it is important to emphasize that the investigation of the properties of the superconducting phase transition is also important to other areas of physics. Here we shall briefly enumerate and discuss several examples.

1.2. Quantum Field Theory and Other Problems

In elementary particle physics the gauge invariant theory similar to the GL theory of superconductivity is called Abelian Higgs Model [8, 9, 10, 11, 12, 13]. The same global $U(1)$ and local $U(1)$ gauge symmetries are present there but the phenomenon of spontaneous symmetry breaking occurs only for imaginary mass of the Higgs ($\psi-$) field, which is the analog of the field ψ, describing the Cooper pair in a superconductor. In the absence of spontaneous symmetry breaking this model would describe an ordinary electrodynamics of charged scalars, but the situation becomes more interesting when the mentioned mass is imaginary and the breaking of symmetry occurs. Now the symmetry breaking phenomena receive other names and another physical interpretation. The breaking of both global and local gauge symmetries ensures a mechanism for a transformation of the two initial scalar fields – analogs of the components ψ' and ψ'' of the complex field $\psi = \psi' + i\psi''$, and two massless photon fields – analogs of the two independent components A_j of the vector potential $\vec{A} = \{A_j; j = 1,2,3; \nabla.\vec{A} = 0\}$ in a superconductor, to four massive particle fields: the so-called Higgs boson, which is analog of the spontaneous order $|\psi| > 0$ in a superconductor, and three massive vector field components, i.e. a massive three dimensional vector field. The mass of this new vector field is proportional to the electric charge $|q|$ and magnitude $|\psi|$ of the Higgs field and, as a matter of fact, this is exactly the way, in which the London penetration length in a superconductor depends on the electron charge $|e|$ and the modulus $|\psi|$ of the superconducting order parameter ψ. Thus the spontaneous breaking of the local gauge symmetry leads to the formation of massive particles without spoiling the gauge invariance of the theory. This is called "the Higgs mechanism". It plays a fundamental role in the unified theory of electromagnetic, weak and strong interactions (see, e.g., Ref. [12, 13]).

It is easy to see that there is something quite common between the phenomena of spontaneous breaking of the continuous symmetries in superconductivity theory and in quantum field theory. The superconducting phase $|\psi| > 0$ is the exact analog of the Higgs boson in the Abelian-Higgs model, whereas the appearance of a massive vector field has its analog in the finite London penetration length, mentioned above. Of course, there is no obstacles in interpreting the phenomena of spontaneous symmetry breaking in quantum field theory as phase transitions by taking the Higgs mode mass as tuning parameter. The phase transition will occur at zero mass of the Higgs boson.

A similar phenomenon of spontaneous breaking of both global $U(1)$ and local gauge symmetries is possible also within the scalar electrodynamics due to mass insertions from the radiation corrections, as shown in Ref. [14]. The radiation corrections, in analogy with the magnetic fluctuations in a superconductor, generate an imaginary mass to the initially massless scalar field in this theory and the latter becomes very similar to the Abelian-Higgs model. Here the symmetry breaking leads to the appearance of massive scalar and vector fields describing neutral scalar meson and vector meson, respectively [14]. The radiation corrections to the Lagrangian of the massless scalar electrodynamics [14] resemble very much, in particular in their mathematical form at $D = 4$, the magnetic fluctuation corrections to the GL free energy of 4D superconductors [7]. The interrelationships between the superconductivity theory and the gauge theories of elementary particles has been comprehensively discussed in Ref. [15]. Note also the relation between the superconductivity GL functional and the CP^{N-1} confinement model (see Ref. [16]) and extensions to non-Abelian theories [17].

The gauge theories, mentioned so far, and their extensions have a wide application in the description of the Early Universe [15, 18]. Another interesting gauge theory is that of the nematic-to-smectic A phase transition in liquid crystals. According to the Kobayashi-McMillan-de Gennes theory [19, 20, 21] the smectic-A order is described by two order parameters: the nematic director vector and the complex scalar which gives the long molecule centers of masses. When a description, analogous to that of superconductors, is introduced, as suggested by de Gennes [21], the director vector is substituted with a gauge vector field, which is quite similar to the vector potential $\vec{A}$ in the GL functional. Apart from some specific features intended to take into account the liquid crystal anisotropy, the effective free energy of the smectic A looks very much like the GL free energy of superconductors (see, also, Refs. [22, 23, 24]). A gauge theory in condensed matter physics that has some (but not very close) similarity with the GL free energy of superconductors is the Chern-Simons-Ginzburg-Landau (CSGL) effective functional of the quantum Hall liquid state, in particular, in the description of phase transitions between the plateaus in the quantum Hall effect [25, 26, 27, 28]. Let us mention also the liquid metallic hydrogen, where the problem of the superconducting-to-superfluid phase transition [29] is also related to the topics, discussed in this review.

1.3. About the Investigation of the Fluctuation-Driven First Order Phase Transition in Superconductors

Now we will focus on how the fluctuations effect the properties of the normal-to-superconducting phase transition in a zero external magnetic field, which is the subject of the present review and in particular cases we shall refer to related topics in other natural systems. We shall consider in more details the HLM effect in 2D and quasi-2D superconductors with a special emphasis on the problem of the theoretical predictions reliability. Following Ref. [6] we will use two theoretical methods.

In Sec. II we use the MF like approximation of Ref. [6], where the spatial fluctuations of superconducting order parameter ψ are neglected, and the magnetic fluctuations are taken into consideration. This treatment was justified for well established type I superconductors, where the London penetration length λ is much smaller than the coherence (correlation) length ξ. Due to the neglecting of spatial fluctuations of ψ, the results of such MF treatment should be valid outside the Ginzburg critical region [4]. A weakly first order phase transition occurs as a result of new small $|\psi|^3$-term, which appears in the effective free energy of 3D superconductors within the framework of the MF like approximation [6]. But this HLM effect was found to be very small and experimentally unobservable even for well established type I bulk (3D-) superconductors, such as Al, where the GL number $\kappa = \lambda/\xi = 10^{-2} \ll 1$. It has been recently shown [30, 31, 32, 33, 34, 35, 36] that the HLM effect is much stronger in quasi-2D superconductors than in bulk (3D) samples. Moreover, as shown in this series of papers, the effect appears by a term of type $|\psi|^2 \ln|\psi|$ in the effective free energy, and essentially depends on the thickness L_0 of quasi-2D superconducting films. These circumstances provide a real opportunity for an experimental verification of the effect – a topic of discussion throughout the present review, and in particular, in Sec. II.

We must emphasize that MF results are not valid for thermodynamic states in the Ginzburg critical region $(\delta T)_G = |T_G - T_{c0}|/T_{c0}$ around the equilibrium phase transition point T_{c0}. In certain classes of high-temperature superconductors the Ginzburg region exceeds 0.1 K whereas in usual low-temperature superconductors it is very narrow, $(\delta T)_G \sim 10^{-12} - 10^{-16}$ K, and the respective critical states are experimentally unaccessible (see, e.g., [4]). If the metastability states, which go along with the weakly first order phase transition predicted by MF, extend over temperature intervals larger than the size of the Ginzburg region, one can conclude that the MF prediction of the HLM effect is reliable. In Sec. II we justify the MF analysis reliability for element superconductors as Al, W, In. We will show that the condition for enough wideness of metastability regions is quite strong and, perhaps, irrelevant in experiments. We will also demonstrate that the magnitude of the critical magnetic field is crucial for HLM effect to be observed in thin superconducting films (see also the discussion in Sec. II.9).

In Sec. III we review some results for the HLM effect obtained with the help of the renormalization group (RG) method [4, 5]. The latter allows simultaneous treatment of both superconducting and magnetic fluctuations in the asymptotically close vicinity of phase transition point. The lack of fixed point of one-loop RG equations for conventional superconductors in a zero external magnetic field was interpreted as a signal for a fluctuation-driven first order phase transition [6]. This result shows that the local gauge magnetic fluctuations are relevant also in the Ginzburg region of strong ψ-fluctuations, and under certain circumstances, they can change the phase transition order to a weakly first order, as is outside the critical region. The RG investigations of the superconducting phase transition order in a zero magnetic field has been recently reviewed in Ref. [37]. While the latter review emphasizes the investigation of the HLM effect in high orders of the loop expansion [4, 5], here we shall stress on RG results concerning the effects of anisotropy, quenched disorder and extended symmetries of the Higgs ψ-field on the phase transition properties. The RG study of unconventional superconductors will also be discussed (Sec. III.7). In the remainder of this paper we will consider mainly the phase transition to the superconducting state but results for other systems and related topics will be also mentioned in brief; see, e.g., Sec. III.3 – III.6.

2. MF Studies

2.1. General GL Functional

The GL free energy [2] of D-dimensional superconductor of volume $V_D = (L_1 ... L_D)$ is given in the form

$$F(\psi, \vec{A}) = \int d^D x \left[a|\psi|^2 + \frac{b}{2}|\psi|^4 + \frac{\hbar^2}{4m}\left| \left(\nabla - \frac{2ie}{\hbar c}\vec{A} \right) \psi \right|^2 + \frac{\vec{B}^2}{8\pi} \right] . \tag{1}$$

In Eq. (1) the first Landau parameter $a = \alpha_0(T - T_{c0})$ is expressed by the critical temperature $T_{c0} = T_c(H = 0)$ in a zero external magnetic field ($H = |\vec{H}|$), $b > 0$ is the second Landau parameter and $e \equiv |e|$ is the electron charge. The square $\vec{B}^2$ of the magnetic induction $\vec{B} = (\vec{H} + 4\pi\vec{M})$, is given by the vector potential $\vec{A}(\vec{x}) = \{A_j(\vec{x}),\ j = 1, ..., D\}$ in the following way

$$\vec{B}^2 = \frac{1}{2} \sum_{i,\, j=1}^{D} \left(\frac{\partial A_j}{\partial x_i} - \frac{\partial A_i}{\partial x_j} \right)^2 , \tag{2}$$

here the vector potential $\vec{A}(\vec{x})$ obeys the Coulomb gauge $\nabla \cdot \vec{A}(\vec{x}) = 0$. For 3D superconductor the relation $\vec{B} = \nabla \times \vec{A}(\vec{x})$ can be used and when $\vec{B} = \vec{B}_0$ is uniform along the z-axis, the Landau gauge $\vec{A}_0(\vec{x}) = B_0(-y/2, -x/2, 0)$ can be applied. This representation can be generalized for $D > 2$ - dimensional systems, where the magnetic induction B_0 is a second rank tensor [38]:

$$B_{0ij} = B_0(\delta_{i1}\delta_{j2} - \delta_{j2}\delta_{i1}). \tag{3}$$

If we use the notation $\vec{x} = (x_1, x_2, \vec{r})$, where $\vec{r}$ is a $(D - 2)$ - dimensional vector, perpendicular to the plane (x_1, x_2), in the $3D$ case we have $\vec{r} = (0, 0, z)$, and

$$B_j = \frac{1}{2}\epsilon_{jkl} B_{0kl} = B_0 \delta_{j3} , \tag{4}$$

where ϵ_{jkl} is the antisymmetric Levi-Civita symbol. The Landau gauge and Eqs. (3) - (4) can be used for uniform $\vec{B} = \vec{B}_0$ when $\delta\vec{B}$ - fluctuations are neglected. In the prevailing part of our study we will apply the general Coulomb gauge of the field $\vec{A}(\vec{x})$, which does not exclude spatial dependent magnetic fluctuations $\delta\vec{B}(\vec{x})$.

In nonmagnetic superconductors, where the mean value $\langle\vec{M}\rangle = (\vec{M} - \delta\vec{M})$ of magnetization $\vec{M}$ is equal to zero in the normal state in a zero external magnetic field, the magnetic induction in presence of external magnetic field takes the form:

$$\vec{B} = \vec{H}_0 + \delta\vec{H}(\vec{x}) + 4\pi\delta\vec{M}(\vec{x}) , \tag{5}$$

where $\vec{H}_0$ is the (uniform) regular part of the external magnetic field and $\delta\vec{H}$ is an irregular part of $\vec{H}$ created by uncontrollable effects. We neglect the irregular part $\delta\vec{H}$ and set $\vec{H}_0 = 0$, then $\vec{B}$ contains only a fluctuation part $\vec{B} \equiv \delta\vec{B}(\vec{x}) = 4\pi\delta\vec{M}(\vec{x})$ that describes the diamagnetic variations of $\vec{M}(\vec{x})$ around the zero value $\langle\vec{M}\rangle = 0$ due to fluctuations

$\delta\psi(\vec{x})$ of the ordering field $\psi(\vec{x})$ above $(T > T_{c0})$ and below $(T < T_{c0})$ the normal-to-superconducting transition at T_{c0}. Note, that the non-fluctuation part $\vec{A}_0 = [\vec{A}(\vec{x}) - \delta\vec{A}(\vec{x})]$ corresponds to the regular part $\vec{B}_0 = (\vec{H}_0 + \langle\vec{M}\rangle) = 0$ of $\vec{B}$ in nonmagnetic superconductors ($\langle\vec{M}\rangle = 0$) in a zero external magnetic field ($\vec{H}_0 = 0$). Then we can set $\vec{A}_0(\vec{x}) = 0$ and, hence, $\delta\vec{A}(\vec{x}) = \vec{A}(\vec{x})$, so we have an entirely fluctuation vector potential $\vec{A}(\vec{x})$, which interacts with the order parameter $\psi(\vec{x})$. This interaction can be of type $|\psi|^2 A$ and $|\psi|^2 A^2$ and generates all effects discussed in the paper.

We accept periodic boundary conditions for the superconductor surface. This means to ignore the surface energy including the additional energy due to the magnetic field penetration in the surface layer of thickness equal to the London penetration depth $\lambda(T) = \lambda_0|t_0|^{-1/2}$, $t_0 = |T - T_{c0}|/T_{c0}$; $\lambda_0 = (mc^2b/8\pi e^2\alpha_0 T_{c0})^{1/2}$ is the "zero-temperature" value of λ. This approximation is adequate for superconductors of thickness $L_0 \gg \lambda(T) \gg a_0$, where a_0 is the lattice constant and $L_0 = min\{L_i, i = 1, ..., D\}$. As we suppose the external magnetic field to be zero $(H_0 = 0)$ or very small in real experiments, the requirement $L_0 \gg \lambda(T)$ can be ignored and we have the simple condition $L_0 \gg a_0$.

In microscopic models of periodic structures the periodic boundary conditions confine the wave vectors $\vec{k}_i = \{k_i = (2\pi n_i/L_i);\ i = 1, ..., D\}$ in the first Brillouin zone $[-(\pi/a_0) \leq k_i < (\pi/a_0)]$ and the expansion of their values beyond this zone can be made either by neglecting the periodicity of the crystal structure or on the basis of assumption that large wave numbers $k = |\vec{k}|$ have a negligible contribution to the calculated quantities. The last argument is widely accepted in the phase transition theory, where the long-wavelength limit $(ka_0 \ll 1)$ can be used. In particular, this argument is valid in the continuum limit $(V_D/a_0^D \to \infty)$. Therefore, for both crystal and nonperiodic structures we can use the cutoff $\Lambda \sim (\pi/a_0)$ and afterwards extend this cutoff to infinity, provided the main contributions in the summations over $\vec{k}$ come from the relatively small wave numbers $(k \ll \Lambda)$. In fact, this is a quasi-macroscopic description based on the GL functional (1), so the microscopic phenomena are excluded from our consideration.

The GL free energy functional takes into account phenomena with characteristic lengths ξ_0 and λ_0 or larger (ξ and λ), where $\lambda(T)$ is the London penetration length, mentioned above, and $\xi(T) = \xi_0|t|^{-1/2}$ is the coherence length [2]; here $\xi_0 = (\hbar^2/4m\alpha_0 T_{c0})^{1/2}$ is the zero-temperature coherence length. In low-temperature superconductors ξ_0 and λ_0 are much bigger than the lattice constant a_0. Having in mind this argument we will assume in our investigation that $\Lambda \ll (\pi/a_0)$. Whether the upper cutoff Λ is chosen to be either $\Lambda \sim 1/\xi_0$ or $\Lambda \sim 1/\lambda_0$ is a problem that has to be solved by additional considerations. According to arguments presented in Ref. [30] and Sec. II.6, we will often make the choice $\lambda \sim \xi_0^{-1}$.

We will use the Fourier expansion

$$A_j(\vec{x}) = \frac{1}{V_D^{1/2}} \sum_k A_j(\vec{k}) e^{i\vec{k}.\vec{x}} \tag{6}$$

and

$$\psi(\vec{x}) = \frac{1}{V_D^{1/2}} \sum_k \psi(\vec{k}) e^{i\vec{k}.\vec{x}} , \tag{7}$$

where the Fourier amplitudes $A_j(\vec{k})$ obey the relation $A_j^*(\vec{k}) = A_j(-\vec{k})$ and $\vec{k}.\vec{A}(\vec{k}) = 0$. The Fourier amplitude $\psi(\vec{k})$ is not equal to $\psi^*(-\vec{k})$ because $\psi(\vec{x})$ is a complex function. For the same reason $\psi(0) \equiv \psi(\vec{k} = 0)$ is a complex number.

The functional (1) is invariant under global $U(1)$ rotations, defined by $\psi(\vec{x}) \rightarrow \psi(\vec{x})\exp(i\alpha)$, where the angle α does not depend on the spatial vector $\vec{x}$, and under the local $U(1)$ gauge transformations $\psi(\vec{x}) \rightarrow \psi(\vec{x})\exp[i\alpha(\vec{x})]$, $\vec{A}(\vec{x}) \rightarrow \vec{A}(\vec{x}) + (\hbar c/2e)\vec{\nabla}\alpha(\vec{x})$. According to the discussion in Sec. I, we have to investigate the spontaneous breaking of these symmetries, that is, the ordered phases and the phase transitions in the superconductor. The HLM effect,on which we are going to focus the attention, is one of the results of the local $U(1)$ symmetry breaking.

2.2. Notes about the MF Like Approximation

The effect of the superconducting fluctuations $\delta\psi(\vec{x})$ on the phase transition properties is very weak, as in usual superconductors, and is restricted in a negligibly small vicinity ($|t_0| \sim 10^{-12} \div 10^{-16}$) of temperature T_{c0}, we will assume that $\delta\psi(\vec{x}) = 0$, i.e., $\psi \approx \langle\psi(\vec{x})\rangle$; from now on we will denote $\langle\psi(\vec{x})\rangle$ by ψ. Therefore, we apply the mean-field approximation with respect to the order parameter $\psi(\vec{x})$. Within this approximation we will take into account the $\delta\vec{A}(\vec{x})$-fluctuations of $\vec{B}_0 = 0$, i.e., $\vec{A}(\vec{x}) = \delta\vec{A}(\vec{x})$. Furthermore, the $\vec{A}(\vec{x})$-fluctuations can be integrated out from the partition function, defined by:

$$Z(\psi) = \int \mathcal{D}A e^{-F(\psi,\vec{A})/k_B T} , \tag{8}$$

where the functional integral $\int \mathcal{D}A$ is given by

$$\int_{-\infty}^{\infty} \prod_{j=1}^{D} \prod_{x \in V_D} dA_j(\vec{x}) \delta[\text{div}\vec{A}(\vec{x})] . \tag{9}$$

The integration is over all possible configurations of the field $\vec{A}(\vec{x})$; the δ-function takes into account the Coulomb gauge.

The partition function $Z(\psi)$ corresponds to an effective free energy $\mathcal{F}_D$ of the D-dimensional system:

$$\mathcal{F}_D = -k_B T \ln Z(\psi). \tag{10}$$

The magnetic fluctuations will be completely taken into account, if only we are able to solve exactly the integral (8). The exact solution can be done for a uniform order parameter ψ. The uniform value of ψ is different from the mean-field value of ψ, because the uniform fluctuations of $\psi(\vec{x})$ always exist, so we should choose one of these two possibilities [30, 35]. The problem of this choice arises after calculating the integral (8) at the next stage of consideration, when the effective free energy $\mathcal{F}_D$ is analyzed and the properties of the superconducting phase ($\psi > 0$) are investigated. The effective free energy is a particular case of the effective thermodynamic potential in the phase transition theory [4, 5] and we must treat the uniform ψ in the way prescribed in the field theory of phase transitions. It will become obvious from the next discussion that we will use a loop-like expansion, which can be exactly summed up to give a logarithmic dependence on $|\psi|$.

Due to the spontaneous symmetry breaking of the global $U(1)$ continuous symmetry of the ground state $\psi \neq 0$, the effective free energies discussed in this Section depend on the modulus $|\psi|$ of the complex number $\psi = |\psi|e^{i\theta}$ but not on the phase angle θ, which remains arbitrary. That is why we will consider the modulus $|\psi|$ as an "effective order parameter" as the angle θ does not play any role in the phenomena investigated in this Section. The quantity $|\psi|$ remains undetermined up to the stage when we determine the equilibrium order parameter $|\psi_0|$ by the equation of state $[\partial\mathcal{F}_D(\psi)/\partial\psi] = 0$. This equation gives the equilibrium value ψ_0 of ψ and the difference $\delta\psi_0 = (\psi_0 - \psi)$ can be treated as the uniform (zero dimensional) fluctuation of field $\psi(\vec{x})$. The $\vec{x}$-dependent fluctuations $\delta\psi(\vec{x})$ have been neglected because of the ψ uniformity. The solution ψ_0 will be stable towards the uniform fluctuation $\delta\psi$, provided the same solution $\psi_0 = |\psi_0|e^{i\theta_0}$ corresponds to a stable (normal or superconducting) phase; the phase angle θ_0 remains unspecified.

We begin our investigation by setting ψ uniform but at some stage we will also ignore the uniform fluctuation $\delta\psi$ and deal only with the equilibrium value ψ_0 of ψ. The equilibrium value will be calculated after taking into account magnetic fluctuations, so it will be different from the usual result $|\psi_0| = (|a|/b)^{1/2}$ [2], where both magnetic and superconducting fluctuations are ignored. This simplest approximation for the equilibrium value of ψ is obtained from the GL free energy (1), provided $e = 0$ and the gradient term is neglected. Hereafter we will keep the symbol $|\psi_0|$ for the equilibrium order parameter in the more general case, when the magnetic fluctuations are not neglected, and will denote the same quantity for $e = 0$ by $\eta \equiv |\psi_0(e = 0)| = (|a|/b)^{1/2}$.

The above described approximation neglects the saddle point solutions of GL equations, where $\langle\psi(\vec{x})\rangle$ is $\vec{x}$-dependent. Therefore, the vortex state that is stable in type II superconductors cannot be achieved. This is consistent with setting the external magnetic field to zero, so the vortex state cannot occur in any type superconductor. Our arguments can be easily verified with the help of GL equations [2] for a zero external magnetic field; the only nonzero solution for ψ in this case is given by $\eta = (|a|/b)^{1/2}$, although the magnetic fluctuations $\vec{A}(\vec{x}) = \delta\vec{A}(\vec{x})$ are properly considered.

In conclusion we can argue that the described method will be convenient for both type I and type II superconductors in a zero external magnetic field, if the ψ-fluctuations have a negligibly small effect on phase transition properties $T_{c0} = T_c(H_0 = 0)$, where T_c denotes the phase transition line for any $H_0 \geq 0$. For type II superconductors in $H_0 > 0$, two lines $T_{c1}(H_0)$ and $T_{c2}(H_0)$ should be defined, usually given by $H_{c1}(T)$ and $H_{c2}(T)$ [2].

2.3. Effective Free Energy

When the order parameter ψ is uniform the functional (1) is reduced to

$$F(\psi, \vec{A}) = F_0(\psi) + F_A(\psi) \tag{11}$$

with

$$F_0(\psi) = V_D(a|\psi|^2 + \frac{b}{2}|\psi|^4) \tag{12}$$

and

$$F_A(\psi) = \frac{1}{8\pi}\int d^D x \left\{ \rho(\psi)\vec{A}^2(\vec{x}) + \frac{1}{2}\sum_{i,j=1}^{D}\left(\frac{\partial A_j}{\partial x_i} - \frac{\partial A_i}{\partial x_j}\right)^2 \right\} . \tag{13}$$

Here $\rho = \rho_0|\psi|^2$ and $\rho_0 = (8\pi e^2/mc^2)$. It is convenient to calculate the partition function $Z(\psi)$ and the effective free energy $\mathcal{F}_{\rm D}(\psi)$ in the $\vec{k}$-space, where Eqs. (9) and (13) take the form

$$\int_{-\infty}^{\infty}\prod_{j=1}^{D}\prod_{\vec{k}>0}^{k\leq\Lambda} d\mathrm{Re}A_j(\vec{k})d\mathrm{Im}A_j(\vec{k})\delta\left[\vec{k}\cdot\vec{A}(\vec{k})\right] \tag{14}$$

and

$$F_A(\psi) = F_A(0) + \Delta F_A(\psi) . \tag{15}$$

Here

$$F_A(0) = \frac{1}{8\pi}\sum_{j,k} k^2 \left|A_j(\vec{k})\right|^2 , \tag{16}$$

and

$$\Delta F_A(\psi) = \rho\sum_{j,k}\left|A_j(\vec{k})\right|^2 ; \tag{17}$$

note, that we have used the Coulomb gauge $\vec{k}.\vec{A}(\vec{k}) = 0$.

Then the partition function (8) becomes

$$\mathcal{Z}(\psi) = e^{-F_0(\psi)/k_B T}\mathcal{Z}_A(\psi) , \tag{18}$$

where

$$\mathcal{Z}_A(\psi) = \int \mathcal{D}A e^{-F_A(\psi)/k_B T} \tag{19}$$

with $F_A(\psi)$ given by (15) and the functional integration defined by the rule (14). With the help of Eqs. (10) - (19) the effective free energy $\mathcal{F}_D(\psi)$ will be

$$\mathcal{F}_D(\psi) = F_0(\psi) + \mathcal{F}_f(\psi) , \tag{20}$$

where $F_0(\psi)$ is given by Eq. (12) and

$$\mathcal{F}_f(\psi) = -k_B T \ln\left[\frac{\mathcal{Z}(\psi)}{\mathcal{Z}(0)}\right] \tag{21}$$

is the ψ-dependent fluctuation part of $\mathcal{F}(\psi)$. In Eq. (20) the ψ-independent fluctuation energy $\{-k_BT\ln\left[\mathcal{Z}_A(0)\right]\}$ has been omitted. This energy should be ascribed to the normal state of the superconductor, which by convention is set equal to zero.

Defining the statistical averages as

$$\langle(...)\rangle = \frac{\int \mathcal{D}\mathcal{A}e^{-F_A(0)/k_BT}(...)}{\mathcal{Z}_A(0)}, \tag{22}$$

we can write Eq. (21) in the form

$$\mathcal{F}_f(\psi) = -k_BT\ln\langle e^{-\Delta F_A(\psi)/k_BT}\rangle. \tag{23}$$

Eq. (23) is a good starting point for the perturbation calculation of $\mathcal{F}_f(\psi)$. We expand the exponent in Eq. (23) and also take into account the effect of the logarithm on the infinite series. In result we obtain

$$\mathcal{F}_f(\psi) = \sum_{l=1}^{\infty}\frac{(-1)^l}{l!(k_BT)^{l-1}}\langle\Delta F_A^l(\psi)\rangle_c\,, \tag{24}$$

where $\langle...\rangle_c$ denotes connected averages [4]. Now we have to calculate averages of type

$$\langle A_\alpha(\vec{k}_1), A_\beta(\vec{k}_2)...A_\gamma(\vec{k}_n)\rangle_c\,. \tag{25}$$

Here we will use the Wick theorem and the correlation function of form

$$G_{ij}^{(A)}(\vec{k},\vec{k}') = \langle A_i(\vec{k})A_j(-\vec{k}')\rangle = \delta_{\vec{k},\vec{k}'}G_{ij}^A(k)\,, \tag{26}$$

where

$$G_{ij}^A(\vec{k}) = \langle A_i(\vec{k})A_j(-\vec{k})\rangle = \frac{4\pi k_BT}{k^2}\left(\delta_{ij}-\hat{k}_i\hat{k}_j\right) \tag{27}$$

and $\hat{k}_i = (k_i/k)$.

The calculation of the lowest order terms $(l = 1, 2, 3)$ in Eq. (24) with the help of (25) - (27) is straightforward. The perturbation terms in (24) are shown by diagrams in Fig. 1. The infinite series (24) can be exactly summed up and the result is the following logarithmic function

$$\mathcal{F}_f(\psi) = \frac{(D-1)}{2}k_BT\sum_k\ln\left[1+\frac{\rho(\psi)}{k^2}\right]\,. \tag{28}$$

The same result for $\mathcal{F}_f(\psi)$ can be obtained by direct calculation of the Gaussian functional integral (8). This is done using the integral representation of δ-function in (9) or (14) but it introduces an additional functional integration that should be carried out after the integration over $A_j(\vec{x})$.

Eqs. (10), (20) and (28) give the effective free energy density

$$f_D(\psi) = \mathcal{F}_D(\psi)/V_D \tag{29}$$

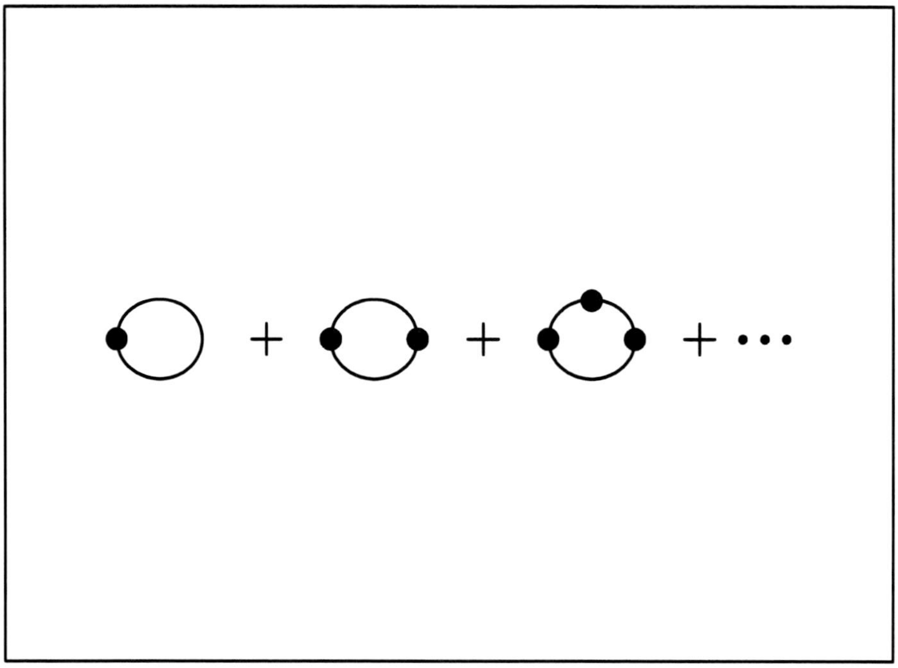

Figure 1. Diagrammatic representation of the series (24); • represents the ρ-vertex in (13) and (17), and the solid lines represent bare correlation functions $\langle |A_j(\vec{k})|^2 \rangle$.

in the form

$$f_D(\psi) = f_0(\psi) + \Delta f_D(\psi) \,, \tag{30}$$

where

$$f_0(\psi) = a|\psi|^2 + \frac{b}{2}|\psi|^4 \tag{31}$$

and

$$\Delta f_D(\psi) = \frac{(D-1)k_BT}{2V_D}\sum_k \ln\left(1 + \frac{\rho}{k^2}\right) \,. \tag{32}$$

Eqs. (20) and (29) - (32) are the basis of our further considerations. We should mention that the fluctuation contribution $\Delta f_D(\psi)$ to $f(\psi)$ transforms to convergent integral in the continuum limit

$$\frac{1}{V_D}\sum_k \to \int \frac{d^Dk}{(2\pi)^D} = K_D \int_0^\Lambda dk.k^{D-1} \,, \tag{33}$$

where $K_D = 2^{1-D}\pi^{-D/2}/\Gamma(D/2)$ for all spatial dimensionalities $D \geq 2$. But the terms in the expansion of the logarithm in (32) are power-type divergent with the exception of several low-order terms in certain dimensionalities D. Therefore, we will work with a finite sum of an infinite series with infinite terms. In our further calculations we will keep the cutoff Λ finite for all relevant terms in $\Delta f_D(\psi)$. This is the condition to obtain correct results.

2.4. 2D-3D Crossover

The dimensional (2D-3D) crossover has been considered in Refs. [34, 39]. Here we follow Ref. [34], where the effective free energy density $f(\psi) \equiv f_3(\psi) = \mathcal{F}_3(\psi)/V_3$ of a thin superconducting film of thickness L_0 and volume $V = V_3 = (L_0 L_1 L_2)$ is derived in a more general way, which allows the investigation of the dimensional crossover. Now one should perform the integration (33) only with respect to the wave vector components k_1 and k_2, corresponding to the large sizes $L_j \gg L_0$ $(j = 1, 2)$. The result for the effective free energy density is [34]:

$$f(\psi) = a|\psi|^2 + \frac{b}{2}|\psi|^4 + k_B T J\left[\rho\left(\psi\right)\right] , \tag{34}$$

where

$$J(\rho) = \int_0^\Lambda \frac{dq}{2\pi} q S\left(q, \rho\right) \tag{35}$$

is given by the sum

$$S = \frac{1}{L_0} \sum_{k_0=-\Lambda_0}^{+\Lambda_0} \ln\left[1 + \frac{\rho(\psi)}{q^2 + k_0^2}\right], \tag{36}$$

and $q = |\vec{q}|, \vec{q} = (k_1, k_2)$.

In Eqs. (34)–(36), the integral $J(\rho)$ and the sum $S(q, \rho)$ over the wave vector $\vec{k} = (\vec{q}, k_0)$ are truncated by the upper cutoffs Λ and Λ_0. The finite cutoff Λ is introduced for the wave number q and Λ_0 stands for k_0.

As our study is based on the quasimacroscopic GL approach the second cutoff Λ_0 should be again related to ξ_0 rather than to the lattice constant a_0, i.e. $\Lambda_0 \sim (1/\xi_0)$, which means that phenomena at distances shorter than ξ_0 are excluded from our consideration. We will assume that the lowest possible value of Λ_0 is (π/ξ_0), as is for Λ, but we will keep in mind that both Λ_0 and Λ can be extended to infinity, provided the main contribution to the integral $J(\rho)$ and the sum S comes from the long wavelength limit $(q\xi_0 \ll 1)$.

In a close vicinity of the phase transition point T_{c0} from normal $(\psi = 0)$ to Meissner state $(|\psi| > 0)$ the parameter $\rho \sim |\psi|^2$ is small and the main contribution to the free energy $f(\psi)$ will be given by the terms in S with small wave vectors $k \ll \Lambda$. This allows an approximate but reliable treatment of the 2D-3D crossover by expanding the summation over k_0 in (36) to infinity - $\Lambda_0 \sim \infty$. A variant of the theory when Λ_0 is kept finite $(\Lambda = \Lambda_0 = \pi/\xi_0)$ can also be developed but the results are too complicated [36]. Performing the summation and the integration in Eqs. (35)–(36) we obtain $J(\rho) = (\Lambda^2/2\pi L_0)I(\rho)$, where

$$I(\rho) = \int_0^1 dy \ln\left[\frac{\text{sh}\left(\frac{1}{2}\, L_0\Lambda\sqrt{\rho + y}\right)}{\text{sh}\left(\frac{1}{2}\, L_0\Lambda\sqrt{y}\right)}\right] , \tag{37}$$

The integral (37) has a logarithmic divergence that corresponds to the infinite contribution of magnetic fluctuations to the free energy of normal phase $(T_{c0} > 0, \varphi = 0)$. Such type of divergence is a common property of many phase transition models. In the present case, as is in other systems, this divergence is irrelevant, because the divergent term does not depend on the order parameter ψ and the free energy $f(\psi)$ is defined as the difference between the total free energies of the superconducting and normal phases: $f(\psi) = (f_S - f_N)$.

Introducing a dimensionless order parameter $\varphi = (\psi/\psi_0)$, where $\psi_0 = (\alpha_0 T_{c0}/b)^{1/2}$ is the value of ψ at $T = 0$, we obtain the free energy (34) in the form

$$f(\varphi) = \frac{H_{c0}^2}{8\pi}\left[2t_0\varphi^2 + \frac{b}{2}|\varphi|^4 + 2(1+t_0)CI(\mu\varphi^2)\right] , \tag{38}$$

with $I(\mu\varphi^2)$, given by Eq. (37), $\mu = (1/\pi\kappa)^2$, $\Lambda = \pi/\xi_0$, and

$$C = \frac{2\pi^2 k_B T_{c0}}{L_0\xi_0^2 H_{c0}^2} . \tag{39}$$

From the equation of state $(\partial f/\partial\varphi = 0)$ we find two possible phases: $\varphi_{00} = 0$ and the superconducting phase $(\varphi_0 > 0)$, defined by the equation

$$t_0 + \varphi_0^2 + \frac{(1+t_0)CL_0\xi_0}{4\pi\lambda_0^2}K(\mu\varphi_0^2) = 0 , \tag{40}$$

where

$$K(z) = \int_0^1 dy\, \frac{\coth\left(\frac{1}{2}\, L_0\Lambda\sqrt{y+z}\right)}{\sqrt{y+z}} . \tag{41}$$

The analysis of the stability condition ($\partial^2 f/\partial\varphi^2 \geq 0$) shows that the normal phase is a minimum of $f(\varphi)$ for $t_0 \geq 0$, whereas the superconducting phase is a minimum of $f(\varphi)$, if

$$1 > \frac{1}{4}(1+t_0)CL_0\Lambda\mu^2\tilde{K}(\mu\varphi_0^2) , \tag{42}$$

where

$$\tilde{K}(z) = \int_0^1 \frac{dy}{y+z}\left[\frac{\coth\left(\frac{1}{2}\, L_0\Lambda\sqrt{y+z}\right)}{\sqrt{y+z}} + \frac{L_0\Lambda}{2\mathrm{sh}^2\left(\frac{1}{2}\, L_0\Lambda\sqrt{y+z}\right)}\right] . \tag{43}$$

The entropy jump $\Delta S = (\Delta S/V) = [-df(\varphi_0)/dT]$ per unit volume at the equilibrium point of phase transition $T_c \neq T_{c0}$ is obtained in the form

$$\Delta_S S(T_c) = -\frac{H_{c0}^2\varphi_{c0}^2}{4\pi T_{c0}}\left[1 + \frac{CI(\varphi_{c0})}{\varphi_{c0}^2}\right] , \tag{44}$$

where $\varphi_{c0} \equiv \varphi_0(T_c)$ is the jump of the dimensionless order parameter at T_c.

The second term in ΔS can be neglected. In fact, taking into account the equation $f[\varphi_0(T_c)] = 0$ for the equilibrium phase transition point T_c we obtain that $|CI(\varphi_0)/\varphi_0^2|$ is approximately equal to $|t_{c0} + \varphi_{c0}^2/2|$, where φ_{c0}^2 and the dimensionless shift of the transition temperature $t_{c0} = t_0(T_c)$ are expected to be much smaller than unity. The latent heat $Q = T_c\Delta S(T_c)$ and the jump of the specific heat capacity at T_c, $\Delta C = T_c(\partial\Delta S/\partial T)$ can be easily calculated with the help of Eq. (44). For this purpose we need the function $\varphi_0(T)$, which cannot be obtained in analytical form from Eq. (40).

Eqs. (38) and (40) can be analyzed numerically. These relatively simple 2D-3D crossover formulae can be used in investigations of specific substances by variation of thickness L_0 of films from $L_0 \gg \Lambda^{-1}$ ($3D$ system) to $a_0 < L_0 \ll \Lambda^{-1}$ (quasi-$2D$ system) and even to a 2D system for $L_0 = a_0$. Then one can vary the effective dimension of the system $D_{eff}(L_0\Lambda)$ as a function of $L_0\Lambda_0$ from $D = 2$ to $D = 3$ [43]. However, from a purely calculational point of view we have found more convenient to consider particular dimensions of interest separately and then to compare the results in order to demonstrate the relevant differences between the bulk ($3D$) and thin film properties. This approach is applied below.

2.5. Effective Free Energy for Particular Dimensions

For purely 2D superconductor, consisting of a single atomic layer, we can use Eqs. (29)-(32) by setting $D = 2$, and calculate $\Delta f_2(\psi)$ with the help of rule (33):

$$\Delta f_2(\psi) = \left(\frac{k_B T}{8\pi}\right)\left[(\Lambda^2 + \rho_0|\psi|^2)\ln\left(1 + \frac{\rho_0|\psi|^2}{\Lambda^2}\right) - \rho_0|\psi|^2 \ln\left(\frac{\rho_0|\psi|^2}{\Lambda^2}\right)\right] . \quad (45)$$

The first term of the above free energy can be expanded in powers of $|\psi|^2$:

$$\Delta f_2(\psi) = \left(\frac{k_B T}{8\pi}\right)\left\{\rho_0|\psi|^2 + \rho_0|\psi|^2 \ln\left(\frac{\Lambda^2}{\rho_0|\psi|^2}\right) + \frac{\rho_0^2|\psi|^4}{2\Lambda^2}\right\} . \quad (46)$$

Thus we obtain the result from Ref. [40]. This case is of special interest because of the logarithmic term in the Landau expansion for $f(\psi)$, but it has no practical application for the lack of ordering in purely 2D superconductors.

For quasi-2D superconductors we assume that $(2\pi/\Lambda) > L_0 \gg a_0$, where L_0 is the thickness of the superconducting film, and the more precise choice of the upper cutoff $\Lambda \ll (1/a_0)$ for the wave numbers k_i is a matter of additional investigation [30]. In order to justify this definition of quasi-2D system one can use the 2D-3D crossover description presented in Sec. 2.4. The summation over the wave number $k_0 = (2\pi n_0/L_0)$ in Eq. (36) cannot be substituted with an integration because $L_0 \ll L_j$ and the dimension L_0 does not obey the conditions, valid for L_j, $(j = 1, 2)$ [41, 42, 43]. Therefore, for such 3D system we must sum over k_0 and integrate over two other components (k_1 and k_2) of the wave vector $\vec{k}$ (see Sec. 2.4). This gives an opportunity for a systematic description of the 2D-3D crossover as shown in Sec. 2.4. In the limiting case of very small thickness the 2D-3D crossover theory (Sec. 2.4) leads to a result, which is obtained more simply in an alternative way, namely, by ignoring all terms corresponding to $k_0 \neq 0$ in the sum in Eq. (32). This corresponds to the supposition that the quasi-2D film thickness cannot exceed $2\pi\Lambda$. Assuming this point of view the real physical size of the quasi-2D film thickness will depend on the choice of the cutoff Λ. It is certain at this stage that $\Lambda \geq \xi_0$, because the (quasi)phenomenological GL theory does not account phenomena of size less than ξ_0. The upper cutoff Λ of wave numbers can be defined in a more precise way at next stages of consideration; see Sec. II.6 - II.8.

For a quasi-2D film we have the expression:

$$\Delta f(\psi) = \frac{2}{L_0}\Delta f_2(\psi)\,, \tag{47}$$

where $\Delta f_2(\psi)$ is given by Eq. (45).

For bulk (3D) superconductor we obtain:

$$\Delta f_3(\psi) = \frac{k_BT}{2\pi}\left[\frac{\Lambda^3}{3}\ln\left(1+\frac{\rho_0|\psi|^2}{\Lambda^2}\right)+\frac{2}{3}\rho_0|\psi|^2\Lambda-\frac{2}{3}\rho_0^{3/2}|\psi|^3\arctan\left(\frac{\Lambda}{\sqrt{\rho_0|\psi|^2}}\right)\right]. \tag{48}$$

The Landau expansion in powers of $|\psi|$ in this form of $f_3(\psi)$ confirms the respective results of Refs. [6, 7], moreover it correctly gives the term of type $\rho_0^2|\psi|^4$, which has been considered small and neglected in these papers.

For 4D-systems $\Delta f_{\rm D}(\psi)$ becomes

$$\Delta f_4(\psi) = \frac{3k_BT}{64\pi^2}\left[\Lambda^2\rho_0|\psi|^2+\Lambda^4\ln\left(1+\frac{\rho_0|\psi|^2}{\Lambda^2}\right)-\rho_0^2|\psi|^4\ln\left(1+\frac{\Lambda^2}{\rho_0|\psi|^2}\right)\right]. \tag{49}$$

The above expression for $\Delta f_4(\psi)$ can be also expanded in powers of $|\psi|$ to show that it contains a term of the type $|\psi|^4\ln\left(\sqrt{\rho_0}|\psi|/\Lambda\right)$, which produces a first order phase transition; this case is considered in the scalar electrodynamics [14], as mentioned in Sec. I. In our further investigation we will focus our attention on 3D and quasi-2D superconductors.

The free energy density $\Delta f_{\rm D}(\psi)$ can be expanded in powers of $|\psi|$ but the Landau expansion can be done only in an incomplete way for even spatial dimensions. Thus $f_2(\psi)$, $f_4(\psi)$, and $f(\psi)$ - the free energy density corresponding to the quasi-2D films, contain logarithmic terms, which should be kept in their original form in the further treatment of the function $\Delta f_{\rm D}(\psi)$ in the Landau expansion. The analysis has been performed [35] in two ways: with and without Landau expansion of $\Delta f_D(\psi)$. These variants of the theory are called "exact" theory (ET) and "Landau" theory (LT), respectively [33]. It has been shown [35] that these two ways of investigation give the same results in all cases except for quasi-2D films with relatively small thicknesses ($L_0 \ll \xi_0$). It seems important to establish the differences between two variants of the theory because the HLM effect is very small and any incorrectness in the theoretical analysis may be a cause for an incorrect result. By same arguments one can investigate the effect of the factor T in $\Delta f_{\rm D}(\psi)$ on the thermodynamics of quasi-2D films. The factor T can be represented as $T = T_{c0}(1+t_0)$ and one may expect that the usual approximation $T \approx T_{c0}$, which is well justified in the Landau theory of phase transitions [2, 4], may be applied. This way of approximation can be performed by neglecting terms in the thermodynamic quantities smaller than the leading ones. On the other hand practical calculations lead to the conclusion that this approximation cannot be made without a preliminary examination because for some quasi-2D films it produces a substantial error of about 10% [35]. LT, in which the factor T is substituted by T_{c0} is referred to as "simplified Landau expansion" - SLT. All three variants of the theory, ET, LT and SLT have been investigated in Ref. [35].

2.6. Limitations of the Theory

The general result (29) - (32) for the effective free energy $f(\psi)$ has the same domain of validity [2] as the GL free energy functional in a zero external magnetic field. When we neglect a sub-nano interval of temperatures near the phase transition point we can use Eq. (1), provided $|t_0| = |T - T_{c0}|/T_{c0} < 1$, or in the particular case of type I superconductors, $|t_0| < \kappa^2$ [2]. Note, that the latter inequality does not appear in the general GL approach. It comes as a condition for the consistency of our approach with the microscopic BCS theory for type I superconductors [2].

Taking the continuum limit we have to assume that all dimensions of the body, including the thickness L_0, are much larger than the characteristic lengths ξ and λ. The exception of this rule is when we consider thin films. Especially for thin films of type I superconductors, where $((2\pi/\Lambda) > L_0 \gg a_0)$, we should have in mind that $\xi(T) > \lambda(T)$, so the inequalities $\xi > \lambda > \xi_0 > \lambda_0$ hold true in the domain of validity of the GL theory $|t_0| < \kappa^2 < 1$. In Ref. [30] a comprehensive choice of the cutoff Λ has been made, namely, $\Lambda = \xi_0$ (the problem for the choice of cutoff Λ is discussed also in Sec. II.7 – II.8). The respective conditions for quasi-2D films of type II superconductors are much weaker and are reduced to the usual requirements: $\kappa > 1/\sqrt{2}$, $|t_0| < 1$ and $(2\pi/\Lambda) > L_0 \gg a_0$.

If we do a Landau expansion of $f_D(\psi)$ in powers of $|\psi|^2$, the condition $\rho \ll \Lambda^2$ should be satisfied. In order to evaluate this condition we substitute $|\psi|^2$ in $\rho = \rho_0|\psi|^2$ with $\eta^2 = |a|/b$, which corresponds to $e = 0$. As $\lambda^2(T) = 1/\rho$, the condition for the validity of the Landau expansion becomes $[\Lambda\lambda(T)]^2 \gg 1$, i. e., $(\Lambda\lambda_0)^2 \gg |t_0|$. Choosing the general form of $\Lambda_\tau = (\pi\tau/\xi_0)$, where τ describes the deviation of Λ_τ from $\Lambda_1 \equiv \Lambda = (\pi/\xi_0)$, we obtain $(\pi\tau\kappa)^2 \gg |t_0|$; $\kappa = (\lambda_0/\xi_0)$ is the GL parameter.

Thus we can conclude that in type II superconductors, where $\kappa = (\lambda_0/\xi_0) > 1/\sqrt{2}$, the condition $(\rho/\Lambda^2) \ll 1$ is satisfied very well for values of the cutoff in the interval between $\Lambda = (\pi/\xi_0)$ and $\Lambda = (\pi/\lambda_0)$, i.e., for $1 < \tau < (1/\kappa)$. For type I superconductors, where $\kappa < 1/\sqrt{2}$ the cutoff value $\Lambda \sim (1/\xi_0)$ leads to the BCS condition $(|t_0| < \kappa^2)$ for the validity of the GL approach. Substantially larger cutoffs ($\Lambda \gg \pi/\xi_0$), for example, $\Lambda \sim (1/\lambda_0)$ for type I superconductors with $\kappa \ll 1$ lead to a contradiction between the BCS condition and the requirement $\rho \ll \Lambda^2$.

In our calculations we often use another parameter $\mu_\tau = (1/\pi\tau\kappa)^2$ and, in particular, $\mu \equiv \mu_1 = (1/\pi\kappa)^2$ and in terms of μ the condition for the validity of $f_D(\psi)$ expansion becomes $\mu|t_0| \ll 1$, or, more generally, $\mu_\tau|t_0| \ll 1$. Choosing $\tau = 1/\pi$, we obtain the BCS criterion for the validity of the GL free energy for type I superconductors [2]. The choice $\tau = (\xi_0/\pi\lambda_0)$ corresponds to the cutoff $\Lambda_\tau = 1/\lambda_0$. As we will see in Sec. II.7 the thermodynamics near the phase transition point in 3D systems has no substantial dependence on the value of the cutoff Λ_τ but it should be chosen in a way that is consistent with the mean-field-like approximation. The cutoff problem for quasi-2D films has been investigated in Ref. [35]. It has been shown that the choice $\Lambda \sim \pi/\xi_0$ is consistent in the framework of GL theory.

Alternatively, the inequality $(\rho/\Lambda^2) \ll 1$ can be investigated with the help of the reduced order parameter φ defined by $\varphi = |\psi|/\eta_0$, where $\eta_0 \equiv \eta(T=0) = (\alpha_0 T_{c0}/b)^{1/2}$ is the so-called zero-temperature value of order parameter for the GL free energy $f_0(\psi)$, given by Eq. (31). The reduced order parameter φ will be equal to $|t_0|$ for $t_0 < 0$, if only the magnetic fluctuations are ignored, i.e., when $|\psi| = \eta$. Using the notation φ, we obtain the condition $(\rho/\Lambda^2) \ll 1$ in the form $\mu_\tau \varphi^2 \ll 1$. This condition seems to be more precise because it takes into account the effect of magnetic fluctuations on the order parameter ψ.

2.7. Bulk Superconductors

2.7.1. Thermodynamics

The effective free energy $f_3(\psi)$ of bulk (3D-) superconductors is given by Eqs. (29) - (31) and (48). The analytical treatment of this free energy can be done by Landau expansion in small $(\sqrt{\rho_0}|\psi|/\Lambda)$. Up to order $|\psi|^6$ we obtain

$$f_3(\psi) \approx a_3|\psi|^2 + \frac{b_3}{2}|\psi|^4 - q_3|\psi|^3 + \frac{c_3}{2}|\psi|^6 , \tag{50}$$

where

$$a_3 = a + \frac{k_B T \Lambda \rho_0}{2\pi^2} , \tag{51}$$

$$b_3 = b + \frac{k_B T \rho_0^2}{2\pi^2 \Lambda} , \tag{52}$$

$$q_3 = \frac{k_B T \rho_0^{3/2}}{6\pi} , \tag{53}$$

and

$$c_3 = -\frac{k_B T \rho_0^3}{6\pi^2 \Lambda^3} . \tag{54}$$

The cutoff Λ in Eqs. (51) - (54) is not specified and can be written in the form $\Lambda_\tau = (\pi\tau/\xi_0)$ as suggested in Sec. 2.6.

It can be shown by both analytical and numerical calculations [31] that $|\psi|^6$-term has no substantial effect on the thermodynamics, described by the free energy (50). That is why we ignore this term. The possible phases $|\psi_0|$ are found as a solution of the equation of state:

$$[\partial f(\psi)/\partial|\psi|]_{\psi_0} = 0 . \tag{55}$$

There always exists a normal phase $|\psi_0| = 0$, which is a minimum of $f_3(\psi)$ for $a_3 > 0$. The possible superconducting phases are given by

$$|\psi_0|_\pm = \frac{3q_3}{4b_3}\left(1 \pm \sqrt{1 - \frac{16a_3 b_3}{9q_3^2}}\right) \geq 0. \tag{56}$$

Having in mind the existence and stability conditions of $|\psi_0|_\pm$-phases [4], we obtain that the $|\psi_0|_+$-phase exists for $(16a_3b_3) \leq 9q_3^2$ and this region of existence always corresponds to a minimum of $f_3(\psi)$. The $|\psi_0|_-$-phase exists for $0 < a_3 < 9q_3^2/16b_3$ and this region of existence always corresponds to a maximum of $f_3(\psi)$, i.e., this phase is absolutely unstable. For $a_3 = 0$, $|\psi_0|_- = 0$ and hence, it coincides with the normal phase. For $9q_3^2 = 16a_3b_3$ we have $|\psi_0|_+ = |\psi_0|_- = 3q_3/4b_3$ and $f_3(|\psi_0|_+) = f_3(|\psi_0|_-) = 27q_3^4/512b_3^3$. Furthermore $f_3(|\psi_0|_-) > 0$ for all allowed values of $|\psi_0|_- > 0$, whereas $f_3(|\psi_0|_+) < 0$ for $a_3 < q_3^2/2b_3$, and $f_3(|\psi_0|_+) > 0$ for $(q_3^2/2b_3) < a_3 < 9q_3^2/16b_3$. The equilibrium temperature $T_{\rm eq}$ of the first order phase transition is defined by the equation $f(|\psi_0|_+) = 0$, which gives the following result:

$$2b_3(T_{\rm eq})a_3(T_{\rm eq}) = q_3^2(T_{\rm eq}) \; . \tag{57}$$

These results are confirmed by numerical calculations of the effective free energy (50) [31].

The equilibrium entropy jump is $\Delta S = V\Delta s$ and $\Delta s = -(df_3(|\psi|)/dT)$ can be calculated with the help of Eq. (50) and the equation of state (55):

$$\Delta s = -|\psi_0|^2\Phi(|\psi_0|) \; , \tag{58}$$

where $\Phi(|\psi_0|)$ is the following function:

$$\Phi(y) = (\alpha_0 + \frac{k_B\Lambda\rho_0}{2\pi^2}) - \frac{\rho_0^{3/2}k_B}{6\pi}y + \left(\frac{k_B\rho_0^2}{4\pi^2\Lambda}\right)y^2 \; . \tag{59}$$

The specific heat capacity per unit volume $\Delta C = T(\partial\Delta s/\partial T)$ is obtained from (58):

$$\Delta C = -\left(\frac{T}{T_{c0}}\right)\frac{\partial|\psi_0|^2}{\partial t_0}\Phi(|\psi_0|) \; . \tag{60}$$

The quantities $\Delta s(T)$ and $\Delta C(T)$ can be evaluated at the equilibrium phase transition point $T_{\rm eq}$, which is found from Eq. (57):

$$\frac{T_{\rm eq}}{T_{c0}} \approx 1 - \frac{k_B\rho_0\Lambda}{2\pi^2\alpha_0} + \frac{\left(\rho_0^{3/2}k_B/6\pi\right)^2}{b + \left(\rho_0^2k_B/2\pi^2\Lambda\right)T_{c0}}\left(\frac{T_{c0}}{\alpha_0}\right) \; , \tag{61}$$

provided $|\Delta T_c| = |T_{c0} - T_{\rm eq}| \ll T_{c0}$. Further we will see that the condition $|\Delta T_c| \ll T_{c0}$ is valid in real substances. The second term in r.h.s. of Eq. (61) is a typical negative fluctuation contribution, whereas the positive third term in r.h.s. of the same equality is typical of first-order transitions [4].

To obtain the jumps Δs and ΔC at $T_{\rm eq}$ we have to put the solution $|\psi_0|_+$, found from Eq. (56) in Eqs. (58) - (60). The result will be:

$$\Delta s = -\frac{q_{3c}^2}{b_{3c}^2}\left\{\alpha_0 + \frac{k_B\rho_0\Lambda}{2\pi^2} - \left(\frac{k_B\rho_0^{3/2}}{6\pi}\right)^2\frac{T_{\rm eq}}{b_{3c}^2}\right\} \; , \tag{62}$$

and

$$\Delta C = \frac{4\alpha_0}{b_{3c}}\left(\alpha_0 T_{c0} - \frac{q_{3c}^2 b}{b_{3c}^2}\right) , \tag{63}$$

where b_{3c} and q_{3c} are the parameters b_3 and q_3 at $T = T_{\rm eq}$. As $|\Delta T_c| = |T_{c0} - T_{\rm eq}| \ll T_{c0}$ we can set $T_{\rm eq} \approx T_{c0}$ in r.h.s. of Eqs. (62) and (63) and obtain $q_{3c} \equiv q_3(T = T_{\rm eq}) \approx q_3(T_{c0})$ and $b_{3c} \approx b_3(T_{c0})$. The latent heat $Q = -T_{\rm eq}\Delta s$ per unit volume of the first order phase transition at $T_{\rm eq}$ can be calculated from Eq. (62).

Now we shall discuss the ratio

$$(\Delta T)_{\rm eq} = \frac{Q}{\Delta C} \tag{64}$$

introduced in Ref. [6] as a measure of the temperature size of the HLM effect – a temperature interval, where the effect can be observed (in [6] the same ratio is denoted by ΔT_1). Keeping only the first terms in Eqs. (62) - (63), we obtain a result for $(\Delta T)_{\rm eq}$ which is four times smaller than that for ΔT_1 [6]. To explain the difference let us mention that Eq. (63) gives the jump ΔC at the equilibrium phase transition point of the first order phase transition, described by $|\psi|^3$ term [4], while the specific heat jump considered in Ref. [6] is equal to the specific heat jump at the standard second order transition $\Delta\tilde{C} = (\alpha_0^2 T_{c0}/b)$ [4], and is four times smaller. In fact, neglecting the second term in Eq. (63) and having in mind that $b_3 \equiv b$, see Eq. (66) below, we have that $\Delta C = 4\Delta\tilde{C}$. Therefore, we obtain that $(\Delta T)_{\rm eq}$ is four times smaller than the respective value ΔT_1 in Ref. [6]. This is valid in an approximation in which the second term in Eq. (63) is neglected (this approximation is justified in Sec. 2.7.2). The approximation of Δs and ΔC by the first (leading) terms in Eqs. (62) and (63) does not change essentially these quantities provided the cutoff Λ is comprehensively chosen (see also Sec. 2.7.2). Note that in Refs. [6, 7] as well as in all preceding papers, the corrections to the leading terms in Δs and ΔC have not been taken into account.

2.7.2. Results for Al

In order to do the numerical estimates we represent the Landau parameters α_0 and b with the help of the zero-temperature coherence length ξ_0 and the zero-temperature critical magnetic field H_{c0}. The connection between them is given by formulae of the standard GL superconductivity theory [2]: $\xi_0^2 = (\hbar^2/4m\alpha_0 T_{c0})$ and $H_{c0}^2 = (4\pi\alpha_0^2 T_{c0}^2/b)$. The expression for the zero-temperature penetration depth $\lambda_0 = (\hbar c/2\sqrt{2}eH_{c0}\xi_0)$ is obtained from the above relation and $\lambda_0 = (b/\alpha_0 T_{c0}\rho_0)^{1/2}$. We will use the experimental values of T_{c0}, H_{c0} and ξ_0 for Al as given in Table 1. The experimental values for T_{c0}, H_{c0} and ξ_0 vary about 10-15%, depending on the method of measurement and the geometry of the samples (bulk material or films) [44] but such deviations do not affect the results of our numerical investigations.

The evaluation of the parameters a_3 and b_3 for Al gives:

$$a_3 = (\alpha_0 T_{c0})\left[t_0 + 0.972 \times 10^{-4}(1 + t_0)\tau\right] , \tag{65}$$

Table 1. Values of T_{c0}, H_{c0}, ξ_0, κ, and $|\psi_0|$ for W, Al, In [44].

Substance	T_{c0} (K)	H_{c0} (Oe)	ξ_0 (μm)	κ	$\lvert\psi_0\rvert \times 10^{-11}$
W	0.015	1.15	37	0.001	0.69
Al	1.19	99.00	1.16	0.010	2.55
In	3.40	281.5	0.44	0.145	2.0

and

$$\frac{b_3}{b} = 1 + \frac{0.117}{\tau} \ . \tag{66}$$

Setting $\tau = 1$ corresponds to the cutoff $\Lambda_1 = (\pi/\xi_0)$ (Sec 2.6). For $\tau = (1/\kappa)_{Al} = 10^2$, which corresponds to the much higher cutoff $\Lambda = (\pi/\lambda_0)$, we have $b_3 \approx b$, i.e., the ρ_0^2 -term in b_3, given by Eq. (52), can be neglected. However, as we can see from Eq. (66), for $\tau = 1$ the same ρ_0^2-correction in the parameter b_3 is of order $0.1b$ and cannot be automatically ignored in all calculations, in contrast to the supposition in Refs. [6, 7]. However, the more important fluctuation contribution in 3D superconductors comes from the $\tau-$term in Eq. (65) for the parameter a_3. This term is of order 10^{-4} for $\tau \sim 1$ and this is consistent with the condition $|t_0| < \kappa^2 \sim 10^{-4}$, but for $\tau \sim 10^2$, i.e., for $\Lambda \sim (\pi/\lambda_0) \sim 10^6\mu$m, the same $\tau-$ term is of order 10^2, which exceeds the temperature interval ($T_{c0} \pm 10^{-4}$) for the validity of BCS condition for Al (Sec. 2.6).

These results demonstrate that for our theory to be consistent, we must choose the cutoff $\Lambda_\tau = (\pi\tau/\xi_0)$, where τ is not a large number ($\tau \rightarrow 1 \div 10$). To be more specific, we set $\Lambda = \Lambda_1 = (\pi/\xi_0)$ as suggested in Ref. [30].

The temperature shift $t_{\rm eq} = t_0(T_{\rm eq})$ for bulk Al can be estimated with the help of Eq. (61). We obtain that this shift is negative and very small: $t_{\rm eq} \sim -10^{-4}$. Note, that the second term in the r.h.s. of Eq. (61) is of order 10^{-4}, provided $\Lambda \sim (1/\xi_0)$ whereas the third term in the r.h.s. of the same equality is of order 10^{-5}. Once again the change of the cutoff Λ to values much higher than (π/ξ_0) will take the system outside the temperature interval, where the BCS condition for Al is valid. Let us note, that in Ref. [31] the parameter t corresponds to our present notation t_0. But the numerical calculation of the free energy function $f_3(\psi)$ in Ref. [31] was made for the SLT variant of the theory and the shifted parameter $(t_0 + 0.972 \times 10^{-4})$ was incorrectly identified with t and this has led to the wrong conclusion for its positiveness at the equilibrium phase transition point $T_{\rm eq}$. As a matter of fact, the shifted parameter $(t_0 + 0.972 \times 10^{-4})$ is positive at $T_{\rm eq}$ but $t_{\rm eq} \equiv t_0(T_{\rm eq})$ is negative, as firstly noted in Ref. [35].

Having in mind these remarks, when we evaluate Δs and ΔC for bulk Al we can use simplified versions of (62) and (63) which means to consider only the first terms in the r.h.s and to take $q_{3c} \approx q_3$ and $b_{3c} \approx b$ at T_{c0}. In this way we obtain

$$Q = -T_{c0}\Delta s = 0.8 \times 10^{-2} \left[\frac{\rm erg}{\rm K\,.\,cm^3}\right] , \tag{67}$$

and

$$\Delta C = 2.62 \times 10^3 \left[\frac{\text{erg}}{\text{cm}^3}\right] . \tag{68}$$

The results are consistent with an evaluation of ΔC for Al as a jump ($\Delta\tilde{C} = \alpha_0^2 T_{c0}/b$) at the second order superconducting transition point [6] that, as we have mentioned above, is four times smaller than the jump ΔC given by Eq. (68).

A complete numerical evaluation of the function $f_3(\psi)$ and the jump of the order parameter at T_{eq} for bulk Al was presented for the first time in Ref. [31]. The results there confirm that the order parameter jump and Q for bulk type I superconductors are very small and can hardly be observed in experiments.

We will finish the presentation of bulk Al with a discussion of the ratio (64). It can be also written in the form

$$(\Delta T)_{\text{eq}} = \frac{32\pi}{9}\left(\frac{k_B^2 T_{c0}^2}{b\alpha_0}\right)\left(\frac{e^2}{mc^2}\right)^3 , \tag{69}$$

and it differs by a factor $1/4$ from the respective result in Ref. [6]. The reason for this difference was explained at the end of Sec. 2.7.1. From Eq. (69) we have

$$(\Delta T)_{\text{eq}} = 6.7 \times 10^{-12}(T_c^3 H_{c0}^2 \xi_0^6), \tag{70}$$

and multiplying the number coefficient in the above expression by four we obtain Eq. (10) presented in Ref. [6].

2.8. Quasi-2D Films

Following Refs. [32, 33] we can present the free energy density $f(\psi) = (F(\psi)/L_1L_2)$ in the form

$$f(\varphi) = \frac{H_{c0}^2}{8\pi}\left[2t_0\varphi^2 + \varphi^4 + C(1+t_0)\Gamma(\mu\varphi^2)\right], \tag{71}$$

where

$$\Gamma(y) = (1+y)\ln(1+y) - y\ln y, \tag{72}$$

To obtain Eqs. (71) - (72) we have set $\Lambda = (\pi/\xi_0)$ and introduced the notation $\varphi = |\psi|/\eta_0$; η_0 is defined in Sec. 2.5. Some of the properties of free energy (71) were analyzed in Ref. [32] for Al films and in Ref. [33] for films of Tungsten (W), Indium (In), and Aluminium (Al). Here we will briefly discuss the main results.

The equilibrium order parameter $\varphi_0 > 0$ corresponding to the Meissner phase, can be easily obtained from the equation $\partial f(\varphi)/\partial\varphi = 0$ and Eq. (71):

$$t_0 + \varphi_0^2 + \frac{C\mu(1+t_0)}{2}\ln\left(1 + \frac{1}{\mu\varphi_0^2}\right) = 0 . \tag{73}$$

The logarithmic divergence in Eq. (73) has no chance to occur because φ_0 is always positive and does not tend to zero.

The largest terms in the entropy jump δs and the specific heat jump δC at the equilibrium first order phase transition point $T_{\rm eq}$ are given by [32, 33]

$$\delta s = -\frac{H_{c0}^2}{4\pi T_{c0}}\varphi_{\rm eq}^2, \tag{74}$$

where $\varphi_{\rm eq} = \varphi_0(T = T_{\rm eq})$, and

$$\delta C = \frac{H_{c0}^2}{4\pi T_{c0}} \,. \tag{75}$$

The latent heat of the phase transition [4] is given by $Q = -T_{\rm eq}\delta s$ and Eqs. (73)-(74). Since the temperatures $T_{\rm eq}$ and T_{c0} have very close values, the difference between the values of Q, δs, and δC at T_{c0} and $T_{\rm eq}$, respectively, can also be ignored, for example, $|\delta C(T_{\rm eq}) - \delta C(T_{c0})|/\delta C(T_{c0}) \ll 1$ and we can use either $\delta C(T_{c0})$ or $\delta C(T_{\rm eq})$ [32].

The equations (71) - (73) corresponding to quasi-2D films are quite different from the respective equations for bulk (3D-) superconductors but it is easily seen that the relatively large value of the order parameter jump φ^2 in thin films again corresponds to relatively small value of the GL parameter κ. That is why we consider element superconductors with small values of κ and study the effect of this parameter, the critical magnetic field H_{c0} and the film thickness L_0 on the properties of the fluctuation-induced first order phase transition.

We use experimental data for T_{c0}, H_{c0}, ξ_0 and κ for W, Al, and In, published in Ref. [44] (see Table 1). In some cases the GL parameter κ can be calculated with the help of the relation $\kappa = (\lambda_0/\xi_0)$ and the available data for ξ_0 and λ_0. In other cases it is more convenient to use the following representation of the zero-temperature penetration depth:

$$\lambda_0 = \frac{\hbar c}{2\sqrt{2}eH_{c0}\xi_0} \,. \tag{76}$$

The value of $|\psi_0|$ in Table 1 is found from

$$|\psi_0| = \left(\frac{m}{\pi\hbar^2}\right)^{1/2}\xi_0 H_{c0} \,. \tag{77}$$

The order parameter dependence on the reduced temperature difference t_0 is shown in Fig. 2 for Al films of different thicknesses. It is readily seen that the behavior of the function $\varphi_0(t_0)$ corresponds to a well established phase transition of first order. The vertical dashed lines in Fig. 2 indicate the respective values of $t_{\rm eq} = t_0(T_{\rm eq})$, at which the equilibrium phase transition occurs, as well as the equilibrium jump $\varphi_0(T_{\rm eq}) = \varphi_{\rm eq}$ for different thicknesses of the film. The parts of the $\varphi_0(t_0)$-curves, which extend up to $t_0 > t_{\rm eq}$, describe the metastable (overheated) Meissner states, which can appear under certain experimental circumstances (see in Fig. 2 the parts of the curves on the r.h.s. of the dashed lines). The value of $\varphi_{\rm eq}$ and the metastable region decrease with the increase of the film thickness, which shows that the first order of the phase transition is better pronounced in thinner films and this confirms a conclusion in Ref. [32].

These results are justified by the behavior of the free energy as a function of t_0. We used Eqs. (71)-(73) for the calculation of the equilibrium free energy $f[\varphi_0(t_0)]$. The free energy

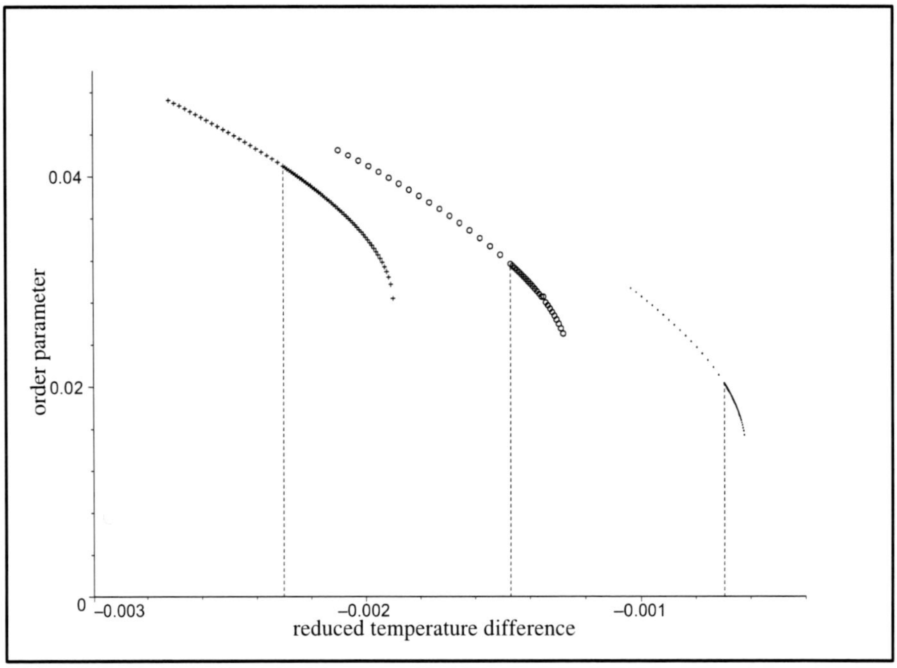

Figure 2. Order parameter profile $\varphi(t_0)$ of Al films of different thicknesses: $L_0 = 0.05$ μm ("+"-line), $L_0 = 0.1$ μm ($\circ$), and $L_0 = 0.3$ μm ($\cdot$) [33].

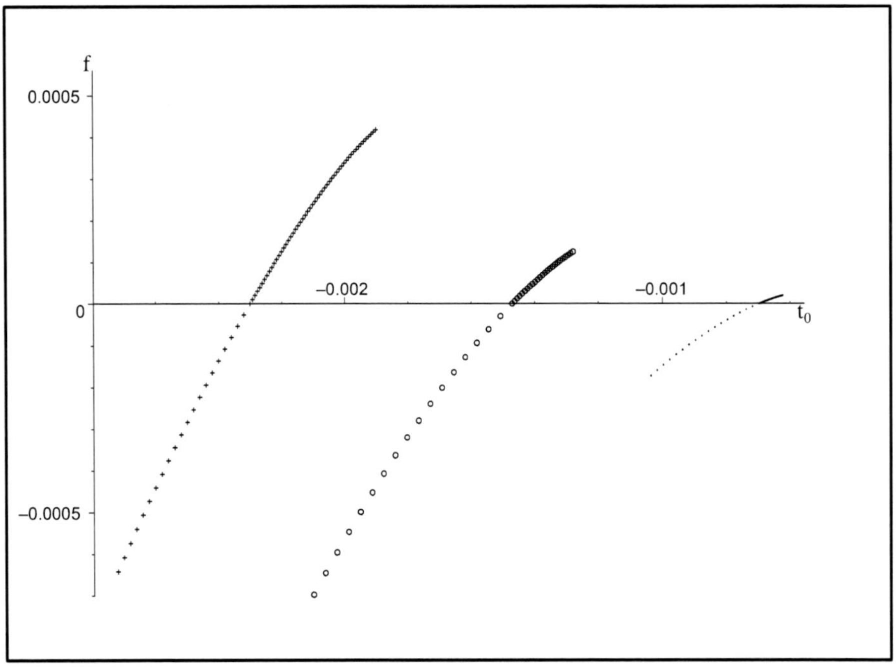

Figure 3. The free energy $f(t_0)$ for Al films of thickness: $L_0 = 0.05$ μm ("+"-line), $L_0 = 0.1$ μm ($\circ$), $L_0 = 0.3$ μm ($\cdot$) [33].

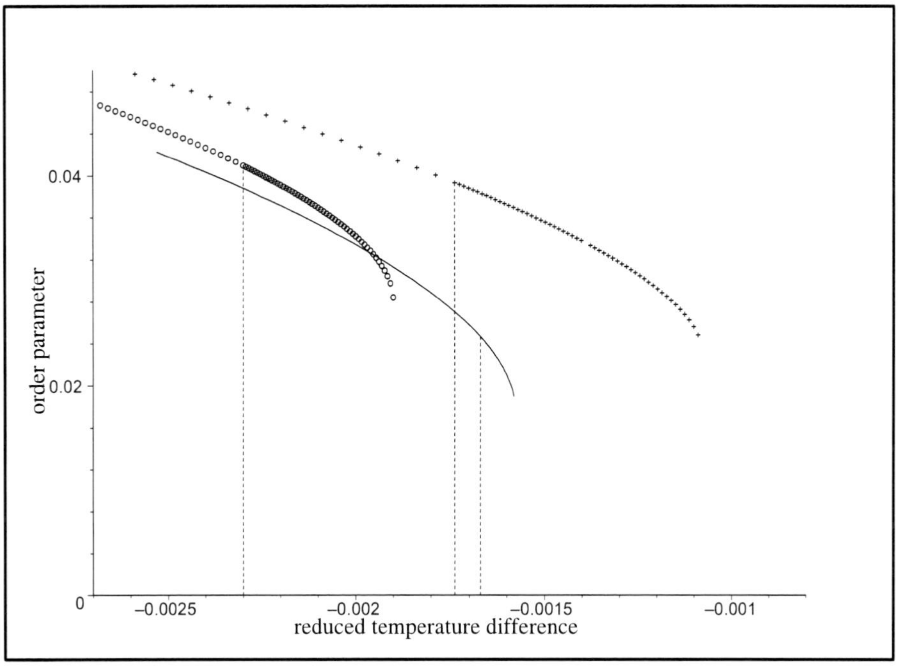

Figure 4. Order parameter profile $\varphi(t_0)$ of films of thickness $L_0 = 0.05\ \mu$m: W ("+"-line), Al ($\circ$), and In ($\cdot$) [33].

for Al films with different thicknesses is shown in Fig. 3. The equilibrium points $T_{\rm eq}$ of the phase transition correspond to the intersection of the $f(\varphi_0)$-curves with the t_0-axis. It is obvious from Fig. 3 that the temperature domain of overheated Meissner states decreases with the increase of the thickness L_0.

The shape of the equilibrium order parameter $\varphi_0(t_0)$ in a broad vicinity of the equilibrium phase transition for thin films ($L_0 = 0.05\mu$m) of W, Al, and In was found from Eq. (73). The result is shown in Fig. 4. The vertical dashed lines in Fig. 4 again indicate the respective values of $t_{\rm eq} = t_0(T_{\rm eq})$, at which the equilibrium phase transition occurs as well as the equilibrium jump $\varphi_0(T_{\rm eq}) = \varphi_{\rm eq}$ in the different superconductors.

The order parameter jump at the phase transition point of In (the In curve is marked by points in Fig. 4) is relatively smaller than for W, and Al, where the GL parameter has much lower values. The same is valid for the metastability domains; see the parts of the curves in Fig. 4 on the left of the vertical dashed lines. It is obvious from Fig. 4 and Table 2 that the equilibrium jump of the reduced order parameter $\varphi_{\rm eq}$ of W has a slightly smaller value than that of Al, although the GL number κ for W has a ten times lower value, compared to κ of Al. Note, that in Fig. 4 we show the jump of $\varphi_{\rm eq}$, but the important quantity is $|\psi|_{\rm eq} = |\psi_0|\varphi_{\rm eq}$. Using the data for $L_0 = 0.05\mu$m from Tables 1 and 2 we find for $|\psi|_{\rm eq}$ the following values: 0.1×10^{11} for Al, 0.05×10^{11} for In, and 0.02×10^{11} for W. This result shows that the value of the critical filed H_{c0} is also important and should be taken into account together with the smallness of GL number when the maximal values of the

Table 2. Values of t_{eq}, φ_{eq}, and Q (erg/cm^3) for films of W, Al, and In with different thicknesses L_0 (μm) [33].

	Al			In			W		
L_0	teq	φ_{eq}	Q	teq	φ_{eq}	Q	teq	φ_{eq}	Q
0.05	−0.00230	0.041	1.95	−0.00167	0.025	3.94	−0.00174	0.039	1.6×10^{-4}
0.1	−0.00147	0.032	0.80	−0.00094	0.017	1.82	−0.00118	0.032	1.1×10^{-4}
0.3	−0.00070	0.023	0.41	−0.00037	0.010	0.63	−0.00064	0.023	5.6×10^{-5}
0.5	−0.00048	0.016	0.20	−0.00029	0.008	0.40	−0.00048	0.020	4.1×10^{-5}
1	−0.00029	0.012	0.11	−0.00013	0.006	0.23	−0.00032	0.016	2.7×10^{-5}
2	−0.00017	0.009	0.06	−0.00008	0.004	0.10	−0.00021	0.013	1.8×10^{-5}

order parameter jump are looked for. Thus the value of the order parameter jump at the fluctuation-induced phase transition is maximal, provided small values of the GL parameter κ are combined with relatively large values of the critical field H_{c0}. In our case Al has the optimal values of these two parameters.

The importance of the zero-temperature critical magnetic field H_{c0} for the enhancement of the jumps of the certain thermodynamic quantities at the equilibrium phase transition point T_{eq} becomes obvious from Eqs. (74), (75) and (77). Eq. (77) shows that the order parameter jump $|\psi|_{eq} = |\psi_0|\varphi_{eq}$ is large for large values of H_{c0} and ξ_0. This is consistent with the requirement for relatively small values of the GL parameter κ, as shown by Eq. (76). Therefore, the unmeasurable ratio $Q/\delta C$ discussed in Ref. [6] does not depend on the value of the critical field H_{c0} but the quantities Q and δC themselves, as well as the order parameter jump $|\psi|_{eq}$ depend essentially on H_{c0}. The values of the reduced order parameter jump φ_{eq} for films of Al, In and W of the same thickness have the same order of magnitude while the respective order parameter jump $|\psi|_{eq}$ is one order of magnitude higher for Al than for W, as shown above. The effect of the critical magnetic field H_{c0} on the latent heat Q is, however, much stronger and, as is seen from Table 2, the latent heat Q in W films is very small and can be neglected while in Al and In films it reaches values, which could be measured in suitable experiments. This is so because the latent heat is proportional to the difference $[H_{c0}^2/8\pi \sim b|\psi_0|^4]$ between the energies of the ground state (superconducting phase at $T = 0$) and the normal state. It is consistent with the fact that the fluctuation contribution to the free energy, i.e., the $C-$term in the r.h.s. of Eq. (71) is generated by the term of type $|\psi|^2 \int d^Dx\vec{A}^2(\vec{x})$ in the GL free energy. At $T = 0$ this free energy term is also proportional to the mentioned difference between the free energies of the superconducting ground state and the normal state.

The shift of the phase transition temperature $t_{eq} = |(T_{eq} - T_{c0})|/T_{c0}$, the reduced value φ_{eq} of the equilibrium order parameter jump $|\psi|_{eq}$, and the latent heat Q of the equilibrium transition are given for films of different thicknesses and substances in Table 2. This table shows that the shift of the phase transition temperature is very small and can be neglected in all calculations and experiments based on them. The values for φ_{eq} for different L_0 and those for $|\psi_0|$ given in Table 1 confirm the conclusion, which we have made for films of Al, In, and W with $L_0 = 0.05\mu$m. The latent heat Q has maximal values for In, where the critical field is the highest for the considered materials.

2.9. Final Remarks

Our analysis shows that the MF studies of the HLM effect have a well defined domain of validity for both 3D- and quasi-2D superconductors. Our conclusion is that the MF theory of the magnetic fluctuations in superconductors and, in particular, the MF prediction for the fluctuation driven weakly first order phase transition in a zero external magnetic field in bulk superconductors [6] and quasi-2D superconducting films [30, 32, 33] is reliable and can be tested by experiments. While the HLM effect in bulk systems is inobservantly small, in quasi-two dimensional superconductors this effect is much stronger and may be observed with available experimental techniques.

Our consideration of quasi-2D superconductors is highly nontrivial in view of the relevance of the effective Landau parameters dependence on the thickness of the films, L_0. We have justified this dependence by simple heuristic arguments and by a reliable consideration of the 2D-3D crossover. In contrast to initial expectations [6] that films made of superconductors with extremely small GL parameter κ such as Al and, in particular, W will be the best candidates for an experimental search of the HLM effect, our careful analysis (firstly published in Ref. [33]) definitely gives somewhat different answer. The Al films still remain a good candidate for transport experiments, through which the jump of the order parameter at the phase transition point could be measured, but surprisingly the W films turn out inconvenient for the same purpose, due to their very low critical field H_{c0}. The importance of the critical magnetic field H_{c0} for the clearly manifested first-order phase transition has been established and discussed, too. Although In has ten times higher GL number κ than Al, the In films can be used on equal footing with the Al films in experiments intended to prove the order parameter jump. Here the choice of one of these materials may depend on other features of experimental convenience. As far as caloric experiments are concerned, the In films seem to be the best candidate due to their high latent heat.

As shown in this review experiments, designed to search the HLM effect, can be performed by both type I and type II superconductors. In experiments the external magnetic field cannot be completely eliminated. Then vortex states may occur for $H = |\vec{H}| > 0$ below $T_c = T_c(H) \leq T_{c0}$ in type II superconducting films and this will obscure the HLM effect. Note, that in both type I and type II superconductors the external magnetic field H generates additional entropy jump at the phase transition point $T_c(H)$ and this effect can hardly be separated from the entropy jump (74) caused by the magnetic fluctuations in the close vicinity of T_{c0}. Therefore, in experiments intended for searching of HLM effect, the external magnetic field should be minimized as much as possible. For quasi-2D superconductors, where the HLM effect is relatively strong and the latent heat can exceed several ergs, one should ensure external fields less than 10 Oe, or, in more reliable experiments, less than 1 Oe [35].

The temperature range around T_{c0}, where the HLM effect is significant, has been estimated in Ref. [6] to a few microdegrees, and later, this estimate has been confirmed in a more accurate way [45]. According to our point of view, the possibilities for the observation of the effect are related mainly with its magnitude, because the range of temperatures where it occurs always exists, even in the critical region of strong ψ-fluctuations, as the RG studies indicate (Sec. III). The review of the results for Al, In, and W in Sec. II shows that the metastability domains (of overheating and overcooling) are much larger than the Ginzburg critical region. This result justifies the reliability of the MF treatment. As one may see from Figs. 2-4 the metastability temperature interval is relatively larger for smaller values of the GL number κ, but in order to ensure large values of some measurable thermodynamic quantities as, for example, the latent heat Q and the specific heat capacity C we must choose a material with large critical field H_c. The experimental verification of the order parameter jump can be made by transport experiments. As we see from Figs. 2-4 and Table 2, this quantity has maximal values for W, where the parameter κ is very small. Having in mind also the large metastability regions in this material, one may conclude that W is a good candidate for a testing the HLM effect by transport experiments (measurements of the superconducting currents), provided some specific disadvantages of this material (quite low T_{c0}, etc.) from the experimental point of view do not contradict to this our conclusion. Another effect, which is relevant to the present discussion, is the known variation of the GL parameter κ with the variation of the thickness L_0 (see, e.g., Ref. [46]. The parameter $\kappa < 1/\sqrt{2}$ of a type-I bulk superconductor may change up to values corresponding to a type-II superconductor with the decrease of the thickness L_0 below 10^{-7}m.

In Ref. [47] a very interesting interrelationship between the first order phase transition and the vortex fluctuations in type-I 2D superconductors has been investigated by both analytical and numerical methods. A recent study [48] of HLM effect in superconducting films has been proven wrong in our comment [49]; see, also, the Corrigendum [50] to the work [48]. In further papers the same authors and co-workers confirm [51, 52] our results, published in Ref. [30, 31, 32, 33, 34, 35], and discussed in this Section. Another paper has been intended to the treatment of gauge effects in superconductors with the help of a Gaussian effective functional [53], which gives results quite near to mean-field ones, and is wrong within the present discussion.

3. RG Studies and Other Topics

3.1. Notes about RG

The RG methods are intended to investigation the strong fluctuation interactions in the critical region of second order phase transitions and multi-critical phenomena [4, 5]. In studies of complex systems with several orderings or with influence of additional gauge fields, as in Eq. (1), usually the $\vec{k}-$space RG methods are used because of their wider applicability. The methods differ from each other in specific technical details but all of them lead to the same predictions about the phase transition properties. The common feature is that the mentioned methods are based on the so-called loop expansion [4, 5, 12]. The RG equations are, in fact, infinite asymptotic series, which are truncated at one-, two-, or, in rare cases, higher orders in the loop expansion. The total infinite series can be summed only in case of trivial (usually exactly solvable) models and in some very exceptional cases as, for example, the RG series for interacting real bosons in the quantum limit $T \to 0$ [54].

For the simple ϕ^4–theory the RG series can be derived and analyzed to high orders in the loop expansion whereas for more complex models this can be practically done within the one- and two-loop approximations. As the RG series are asymptotic, normally, the one- and two-loop orders give all important features of the specific system of interest: (i) the presence of stable fixed points (FP) of RG equations and the associated with them types of critical behavior, and (ii) the lack of stable FPs and related conclusions about the lack of standard critical or multi-critical behavior. The higher orders of the theory are more relevant to investigations of the asymptotic properties of the loop series than for obtaining of qualitatively new critical behavior, or, other new qualitative characteristics of the system. The latter are reliably obtained within the one- and two-loop approximations.

In our consideration we shall give examples of RG equations in the original variant of the Wilson-Fisher RG approach; see, e.g., Ref. [4]. It is based on the loop expansion and an explicit scaling procedure for the wave vectors ($\vec{k} = e^{-l}\vec{k}'$) and the fields $\psi(\vec{k}) = e^{(1-\eta_\psi/2)l}\psi'(\vec{k}')$ and $A_j(\vec{k}) = e^{(1-\eta_A/2)l}A'_j(\vec{k}')$, where $e^l > 1$ is the so-called re-scaling factor ($l > 0$), and η_ψ and η_A are the anomalous dimensions (*alias* Fisher exponents) of the fields $\psi(\vec{k})$ and $A_j(\vec{k})$, respectively.

3.2. The Order of the Phase Transition

In Ref. [6] the simultaneous effect of ψ- and $\vec{A}$–fluctuations in the fluctuation Hamiltonian (1) has been investigated by one-loop RG and $\epsilon = (4-d)$-expansion [4] for the general case of $n/2$-component complex vector field $\psi = \{\psi_\alpha; \alpha = 1, ..., n/2\}$; note, that this field is equivalent to a n-component real vector field. It has been shown [6] that a stable FP exists below four dimensions $d = (4-\epsilon)$ only for symmetry indices $n \geq 365.9$, which are far above the real symmetry index $n = 2$ for usual superconductors and numbers $n = 4, 6$ corresponding to superconductors with certain unconventional Cooper pairings [55]. For $n < 365.9$ real FP does not exist at all. Besides, the $1/n$-expansion has been used for the calculation [6] of the critical exponents in the so-called large-n limit (alias "Hartree limit") [4, 5], where ($n > 365.9$) the phase transition is definitely of second order.

According to the normal interpretation of RG results, the lack of stable FP is an indication for a lack of standard second order phase transition. But usually, additional (non-RG) arguments are needed to determine the actual order of the phase transition. In the present case, the lack of stable FP could be a result of the same mechanism that produces a fluctuation-driven weakly first order phase transition in MF approximation (Sec. 2). As

RG takes into account both fluctuations of ψ and $\vec{A}$, one can conclude that the result for the weakly-first order phase transition in a zero external magnetic field is valid for both type I and II superconductors. Of course, the size of the effect (the size of jumps of energy, latent heat, order parameter) will depend on the specific substance. This path of investigations has been further developed in the paper [7] and several problems, opened in the short Ref. [6], have been solved.

In Ref. [40], where both 2D and 3D superconductors are considered, the HLM effect has been confirmed in one-loop order (annealed disorder [40]) as well as the availability of second order phase transition for large-n has been proven in another variant of the Hartree limit ($n \to \infty$).

Next, the effect has been confirmed in a RG investigation of the equation of state below the phase transition point T_{c0} [56, 57]. These investigations show that the effect may occur under quite restricted conditions for the vertex parameters e and b of the functional (1), in particular, when the effective charge $|e^*|$ exceeds some value. A strong restriction on the HLM effect, but in the opposite direction - a requirement for a sufficiently small effective electric charge $|e^*|$, has been found also by a Monte Carlo (MC) study [58] of a lattice version of the model (1). Another MC study [59] concludes that at fixed effective charge ($|e^2| = 5$) the HLM effect strongly depends on the GL parameter $\kappa = \lambda/\xi$: it is well established for $\kappa \ll 1$, becomes very weak for $\kappa \sim 1$ and vanishes for large κ. On the basis of this picture, a proposal is made [59] that a tricritical point exists at some "tricritical" value of κ. As κ is related with the charge e, this proposal seems to be in conformity with other investigations. Note, that the first prediction for change of the phase transition order with the variation of κ and for the possibility of "tricritical point" to occur at some value of κ, has been made in Ref. [60] on the basis of duality arguments [61] and analytical calculations; see further development of this approach in Refs. [62, 63, 64, 65]. A further MC study [66] of type II superconductors indicates that these systems undergo a second order phase transition of universality class 3D *XY* rather than a weakly first order phase transition. The two-loop RG studies [67, 68] show that the account of higher orders of the loop expansion allows a lowering of the critical value $n_c = 365.9$, but the RG equations still have no real FP at $D = 3$ and $n = 2$ (conventional 3D superconductors), except for a treatment by Padè analysis [68] (see, also, the review [37]). In $2 + \epsilon$ dimensions, the nonlinear σ model exhibits a second-order phase transition for all values of $n > 0$ [69]. Owing to this result and more extensive studies [71, 72] of the dependence of the effective free energy on the GL parameter κ, one may suppose that the critical value n_c vanishes at some dimension $2 < d < 4$, that is, that a tricritical point "located" at some specific value of κ exist where the superconducting phase transition changes its order; see also, Ref. [70]. RG studies [72, 73] at fixed dimension d have also been carried out with the aim to determine the phase transition order (see, also. Ref. [70] for a comment).

3.3. Disorder Effects

3.3.1. Quenched Impurities

Here we will discuss mainly disorder described by quenched impurities and inhomogeneities with Gaussian distributions (see, e.g., Ref. [4]). The effects of annealed and quenched disorder in classical versions of Abelian-Higgs models, equivalent to the GL model (1) have been investigated by Hertz [74, 75] in the context of the theory of spin glasses in case of Dzyaloshinskii-Moriya interaction. The specific feature of this approach is that the disorder is associated with the vector gauge field $\vec{A}(\vec{x})$ and can be used to describe superconductors in random, uncorrelated, or, which is the same, short-range (δ-) correlated magnetic fields rather than usual ones. In high energy physics this approach makes contact with gauge fields, where a Higgs field is coupled to a random color field. It is natural to expect, as has been proven [40, 74], that the annealed gauge model will bring a fluctuation-driven first order phase transition at usual symmetry indices and a second order phase transition in the Hartree limit ($n \to \infty$). In Ref. [40] the phase transition in case of quenched impurities has been predicted to be of second order in a calculation to one-loop order, in contradiction to the conclusion for a lack of stable nontrivial FP of the RG equations within the same one-loop order [74]. The origin of this discrepancy is clear from the argument that the lack of stable FP may result without any change of the Hamiltonian structure, whereas the conclusion for the second order of phase transition in Ref. [40] has been made, perhaps, incorrectly, based on the argument of the $|\psi|^4$-term absence for $D = 3$ and the $|\psi|^2\ln|\psi|$-term absence for $D = 2$.

More usual case of quenched disorder in Abelian-Higgs models has been considered in Refs. [76, 77]. This is the case when the parameter a in (1) has a random part $\varphi(\vec{x})$, i.e. a is substituted with $\tilde{a} = a + \varphi(\vec{x})$, where the Fourier amplitudes $\delta(\vec{k})$ of the *random* function $\varphi(\vec{x})$ obey, for example, the Gaussian distribution

$$\langle \varphi(\vec{k})\varphi(\vec{k}')\rangle_{rand} = \Delta(k)\delta_{-k,k'}. \tag{78}$$

Here the parameter $\Delta(k) \geq 0$ is a non-negative function for any $0 \leq k \leq \Lambda$ (hereafter, $\Lambda = 1$). This disorder is of type random impurities, or, what is the same, "random critical temperature" [4]. In case of the so-called "uncorrelated", or, (δ-) short range correlated quenched impurities [$\Delta(k) \equiv \Delta = const$], the system exhibits a spectacular competition between the impurities and gauge effects. We shall show this by the example of the RG equations derived in Ref. [77]. Using units, in which $k_B = T_{c0} = \Lambda = 1$ and the notations $q_0 = (2e/\hbar c)$ and $\gamma = \hbar^2/4m$ the RG equations corresponding to the impurity model given by Eqs. (1) and (78) can be written in the form

$$a' = e^{(2-\eta_\psi)l}\left[a + \frac{1}{8\pi^2}\left(\frac{n+2}{2}b - \Delta\right)\left(\frac{1-e^{-2l}}{2\gamma} - \frac{a}{\gamma^2}l\right) + \frac{3}{4\pi}\left(1 - e^{-2l}\right)\gamma q_0^2\right], \tag{79}$$

$$b' = e^{(\epsilon-2\eta_\psi)l}\left\{b + \frac{b}{8\pi^2\gamma^2}\left[6\Delta - \frac{1}{2}(n+8)b\right]l - 12\left(\gamma q_0^2\right)^2 l\right\}, \tag{80}$$

$$\Delta' = e^{(\epsilon - 2\eta_\psi)l}\left\{\Delta + \frac{\Delta}{8\pi^2\gamma^2}\left[4\Delta - (n+2)b\right]l\right\}, \tag{81}$$

$$\gamma' = e^{-\eta_\psi l}\gamma\left[1 - \frac{3}{2\pi}q_0^2 l\right], \tag{82}$$

$$e^{\eta_A l} = 1 + \frac{n}{12\pi}q_0^2 l, \quad q_0' = e^{(\epsilon - \eta_A)l/2}q_0. \tag{83}$$

When the disorder is neglected ($\Delta = 0$; pure superconductors) the Eqs. (79) - (83) coincide with those for pure superconductors [6] and give the results briefly discussed in Sec. 3.2. For impure superconductors ($\Delta > 0$), these equations describe a new stable FP of focal type [77] for dimensions $D < 4$, which exists for symmetry indices $n > 1$ and has a physical meaning at dimensions $D > D_c(n) = 2(n+36)/(n+18)$ [77]. In the impure superconductor, the new focal FP occurs exactly in the domain of symmetry indices $n < 365.9$, where the HLM analysis [6] yields a lack of FP for the respective pure system (for details, see Ref. [77]). Having in mind the asymptotic nature of the ϵ-expansions within the RG approach [4, 78], one may conclude that this focal FP governs the critical behavior at the real dimension $D = 3$ in conventional superconductors ($n = 2$), although the direct substitution of $n = 2$ in the above expression for the "lower critical" dimension $D_c(n)$ yields $\mathrm{D}_c(2) = 3.8$. This problem has been further investigated in Ref. [79]. Long-range quenched disorder within the same RG approach to the model (1) has been considered in Ref. [80].

3.3.2. Random Field Disorder

The effect of another relevant type of *quenched* disorder – the so-called "random field disorder" [4] – on the phase transitions described by the GL functional (1) has been considered in Ref. [81]. In order to describe this type of disorder one should add to the GL model (1) the following terms [81]:

$$-\int d^D x \left[h^*(\vec{x}).\psi(\vec{x}) + c.c.\right], \tag{84}$$

where $h = \{h_\alpha; \alpha = 1, \ldots, n/2\}$ is a random field with a Gaussian distribution of the type

$$\langle h_\alpha^*(\vec{k})h_\beta(\vec{k}')\rangle_{rand} = \delta_{\alpha\beta}\delta_{\vec{k}\vec{k}'}h_0^2 k^{-\theta}. \tag{85}$$

In Eq. (85), $\theta = 0$ describes the $k \sim 0$ asymptote of short-range random correlations (in $\vec{x}$-space, $|\vec{x} - \vec{x'}|^{a-d}$; $a \leq 0$), and $0 \leq \theta \leq d$ describes long-range correlations ($0 \leq \theta = a < d$); see Refs. [4, 82, 83].

In both short-range and long-range random field correlations, a new stable FP has been obtained [81]. For short-range random correlations this FP is stable below the upper critical dimension $D_U = 6$ and for symmetry indices $n > n_c = 10$. This FP describes a quite specific critical behavior. The situation for long range random correlations is more complicated and here we will advise the reader to follow the discussion in Ref. [81].

3.3.3. General Conclusion

The annealed disorder leaves the phase transition properties almost identical to those in the pure Abelian-Higgs model (1), whereas the quenched disorder gives a lower critical values of the symmetry index n, below which the weakly first order phase transition (HLM effect) exists. Thus one may conclude that the quenched disorder acts in a direction of "smearing" the fluctuation-driven first order phase transition, i.e. the HLM effect seems to be much weaker in such systems.

3.4. Liquid Crystals

The first theoretical study of the HLM effect on the properties of the nematic-smectic A phase transition in liquid crystals has been performed in Refs. [22, 23]. The main result of these investigations is the confirmation of the HLM effect in a close vicinity of this liquid crystal phase transition. Perhaps, because of the liquid crystal anisotropy, which explicitly enters in the propagator of the gauge field [23], here the critical value $n_c = 38.17$ of the symmetry index n, below which the weakly first order transition occurs, is lower than in pure superconductors ($n_c = 364.9$) but it is still quite above the real value $n = 2$. Thus the RG predictions in Refs. [22, 23] in seventies of the previous century were in favor of a weakly first order phase transition whereas the experimentalists [84, 85] at the same time claimed that the same nematic-smectic A phase transition is of second order (for triple and tricritical points in nematic-smectic-A-smectic-C systems, and the interrelationship to the present problem, see Ref. [86]), as well as later papers [87, 88]. While in 3D superconductors the HLM effect is very weak and unobservable with available experimental techniques, the size of the same effect in 3D liquid crystals allows an observation in case of very precise experiment and elimination of obscuring effects as, for example, neighborhood to tricriticality and liquid crystal anisotropy. Experiments [89, 90] consistent with the HLM effect has been carried out but, as noticed in Ref [91], these experiments are at their resolution limits and the full implications of the HLM theory remains to be tested. On the other hand, experiments in Ref. [91] demonstrate a discrepancy with the simple approximation (ψ =const) for the smectic-A filed within the mean-field like approach (Sec. II).

3.5. Quantum Effects

The first account of quantum correlations ("fluctuations") [4, 92, 93] on the properties of the phase transition to superconducting state has been performed in Ref. [94]. It has been shown [94] that the dynamical critical exponent z, produced by the quantum correlations at finite temperatures ($T > 0$), is given by $z = 2 + 18\epsilon/n$. This is purely gauge result because, as shown in Ref. [95], the critical dynamics of a superconductor when neglecting the local gauge effects ($\vec{A} \equiv 0$) is identical to that described by the time dependent GL model (TDGL) in case of lack of conservation laws [96]; a prediction of the latter result has been given for the first time in Ref. [92].

Another paper [97] on a gauge model of type (1) is intended to the investigation of the quantum phase transition ($T_{c0} = 0$) in granular superconductors. The simultaneous effect of the local $U(1)$ symmetry, disorder and quantum fluctuations has been considered in the Ref. [98, 99], where disordered electronic systems at $T = 0$ are studied. The problem was further extended to the treatment of quantum phase transitions in underdoped high-temperature superconductors [100, 101].

In Ref. [102] the quantum phase transition at zero temperature from superconducting state to the state of normal metal has been discussed within the framework of the HLM approximation (Sec. 2.2). In the limit $T \rightarrow 0$, a dimensional crossover of type $d \rightarrow (d+1)$ occurs due to the quantum fluctuations of the gauge field $\vec{A}(\tau, \vec{x})$, where τ is the imaginary time [4, 93]. Thus the effective free energy of the three dimensional "quantum system" at $T = 0$ corresponds to the free energy of the respective four-dimensional (4D) classical system. Remember, that the latter is described by Eqs. (29)-(31) and (49). Thus the MF like analysis of 3D superconductor in the zero-temperature limit $T \rightarrow 0$ will give a fluctuation-driven first order phase transition, described by a logarithmic term of type $|\psi|^4 \ln|\psi|$ rather than the $|\psi|^3$-term known from Ref. [6]. Accordingly, in the same zero-temperature limit ($T_c = 0$) the 2D- and quasi-2D superconductors will be described by an effective free energy, containing a $|\psi|^3$-term as is in 3D superconductors with $T_c \neq 0$. The complete analysis of the quantum phase transitions ($T_c = 0$) of type-II superconductors beyond the MF-like approximation (Sec. 2.2) requires the account of both thermal and quantum fluctuations of the order parameter field $\psi(\tau, \vec{x})$ with the help of RG; for example, such fluctuations have been systematically taken into account in Refs. [97, 98].

3.6. Complex Systems

The RG investigation [103] of systems with superconducting and other (non-magnetic) orderings shows that the gauge field $\vec{A}(\vec{x})$ leads to a drastic modification of the critical behavior in a close vicinity of bicritical and tetracritical points (for such points, see, e.g., Ref. [2, 4]). Near these multicritical points the superconducting fluctuations can be enhanced by the fluctuations of another ordering field and, hence, the HLM effect in such complex systems seems to be stronger than in the more simple case, discussed so far [104]. The coupling between the superconducting field ψ and the magnetization mode $\vec{M}(\vec{x})$ in models of ferromagnetic superconductors intended to describe the coexistence of superconductivity and ferromagnetism in ternary compounds, has been investigated by RG in Ref. [105, 106]. It has been shown again that the gauge field $\vec{A}(\vec{x})$ produces a weakly first order phase transition, enhanced by the M-fluctuations. The same problem has been further opened in Ref. [107] by a lattice version of the GL model and dual arguments [61] with the conclusion that under certain circumstances the phase transition from the disordered phase to the phase of coexistence of ferromagnetism and superconductivity should be, at least in type II superconductors, of second order in contrast to the RG prediction [105, 106]. Another investigation of coupled magnetic and superconducting order parameters, where the gauge effects are also relevant, is intended to describe a state of vortex solid in rear-earth and borocarbide compounds [108, 109].

3.7. Unconventional Superconductors

Another interesting problem is the behavior of unconventional superconductors [55, 110], where the field ψ is a complex vector: $\psi = \{\psi_\alpha, \alpha = 2, 3\}$, i.e., $n = 4$ or 6, depending on the type of the unconventional Cooper pairing. As already mentioned, in RG studies we may consider an arbitrary value of the symmetry index $n/2 \geq 1$, and to put $n/2 = 2, 3$ in our discussion of real unconventional superconductors. In order to treat these systems we must add two new terms to the free energy (1), namely,

$$\int d^D x(\frac{u}{2}|\psi^2|^2 + \frac{v}{2}\sum_{\alpha=1}^{n/2}|\psi_\alpha|^4), \tag{86}$$

where the parameter u describes the anisotropy of the unconventional (non- s-wave) Cooper pairing and the v-term has been included to represent the crystal anisotropy; as given in (86) this term represents a cubic crystal anisotropy. It has been shown [111], that HLM FP in this case is unstable towards the parameters u and v, describing the anisotropy of the Cooper pair and the crystal anisotropy, and the new FP points that appear in the theory are unstable for usual values of the symmetry index ($n = 4, 6$). Analytical calculations [111] have been used to demonstrate that one of the FP points that appear in the RG equations, exists for $n > n_c = 10988$ and is stable for some values above n_c, at least, in the Hartree limit ($n \rightarrow \infty$). Owing to these results, and mainly due to the very high value of n_c, a conclusion has been made that the HLM effect should be more pronounced in unconventional superconductors [111].

A next step along this path of RG studies has been performed in Ref. [112], where unconventional superconductors with quenched impurities have been investigated by RG to one-loop order. For the most interesting cases of $n = 4$ and $n = 6$, the RG equations have a focal FP, which is known from the results for conventional ($u = v \equiv 0$) impure superconductors (Sec. 3.3.1). However, here this FP is unstable towards the parameters u and v, which means that the superconducting phase transition in a close (asymptotic) vicinity of the phase transition point is not a conventional second order phase transition but rather this transition is a modified version of the fluctuation-driven first order phase transition, known from Ref. [6].

Another interesting problem is related with the gauge effects in ferromagnetic unconventional superconductors such as UGe_2, URhGe, $ZrZn_2$, where a rich picture of various phase transitions and (multi)critical points occurs [113], including the existence of a completely new type of finite temperature critical behavior [114, 115, 116] and quantum critical points. This problem is quite new and since now there are no investigations of the effects of the gauge field fluctuations $\delta\vec{A}(\vec{x})$ on the phase transition properties in ferromagnetic unconventional superconductors.

3.8. External Field Effect

In real experiments the external magnetic field can hardly be completely eliminated. The study [38] of the HLM effect in the wider context of nonzero external magnetic field H_0, shows that the weakly first phase transition, discussed so far can be obtained in Landau gauge from a renormalized theory of the equation of state in one loop order; see also Ref. [45]. The phase transition at the second critical magnetic field H_{c2} has been studied by a $\epsilon = (6 - d)$ expansion within the one loop approximation with the conclusion that the magnetic fluctuation effects should produce a fluctuation induced first order phase transition [117]. Further investigations of this problem have been performed in the Hartree limit $(n \to 0)$ [118, 119], and in higher orders of the loop expansion [120, 121]; for comments, see Refs. [122, 123].

Acknowledgements

Financial support by Grants No. Phys.1507 (NFSR, Sofia). D.I.U. thanks the hospitality of MPI-PKS (Dresden), where his work on this article was performed.

References

[1] V. L. Ginzburg, L. D. Landau, *Zh. Eksp. Teor. Fiz.* **20**, 1064 (1950)

[2] E. M. Lifshitz and L. P. Pitaevskii, *Statistical Physics*, Part 2, [Landau and Lifshitz Course of Theoretical Physics, vol. 9] (Pergamon Press, Oxford, 1980).

[3] A. A. Abrikosov, *Zh. Eksp. Teor. Fiz.* **32**, 1442 (1957) [*Sov. Phys. JETP* **5**, 1174 (1957)].

[4] D. I. Uzunov, *Theory of Critical Phenomena* (World Scientific, Singapore, 1993).

[5] J. Zinn-Justin, *Quantum Field Theory and Critical Phenomena* (Clarendon Press, Oxford, 1993).

[6] B. I. Halperin, T. C. Lubensky, and S. K. Ma, *Phys. Rev. Lett.* **32**, 292 (1974).

[7] J-H. Chen, T. C. Lubensky, and D. R. Nelson, *Phys. Rev. B* **17**, 4274 (1978).

[8] P. W. Higgs, *Phys. Lett.* **12**, 132 (1964); *Phys. Rev. Lett.* **13**, 508 (1964); *Phys. Rev.* **145**, 1156 (1966).

[9] G. S. Guralnik, C. R. Hagen, and T. W. B. Kibble, *Phys. Rev. Lett.* **13**, 585 (1964).

[10] F. Englert, R. Brout, *Phys. Rev. Lett.* **13**, 321 (1964).

[11] T. W. B. Kibble, *Phys. Rev.* **1955**, 1554 (1967).

[12] S. Coleman, *Aspects of Symmetry* (Cambridge University Press, Cambridge, 1985).

[13] M. Guidry, *Gauge Theories* (Wiley-VCH Verlag, Weinheim, 2004).

[14] S. Coleman and E. Weinberg, *Phys. Rev. D* **7** (1973) 1888.

[15] A. D. Linde, *Rep. Progr. Phys.* **42**, 389 (1979).

[16] S. Hikami, *Progr. Theor. Phys.* **62**, 226 (1979).

[17] E. Brézin, C. Itzykson, J. Zinn-Justin, and J-B. Zuber, *Phys. Lett. B* **82**, 442 (1979).

[18] A.Vilenkin, and E.P.S.Shellard, *Cosmic Strings and Other Topological Defects*, (Cambridge Univesity Press, Cambridge, 1994), Ch. 2.

[19] K. K. Kobayashi, *Phys. Lett. A* **31**, 125 (1970; *J. Phys. Soc. Jpn.*, **29**, 101 (1970).

[20] W. L. McMillan, *Phys. Rev. A* **4**, 1238 (1971); **6**, 936 (1972).

[21] P. G. de Gennes, *Solid State Commun.*, **10**, 753 (1972); *Mol. Cryst. Liq. Cryst.* **21**, 49 (1973).

[22] B. I. Halperin, T. C. Lubensky, *Solid State Commun.* **14**, 994 (1974).

[23] T. C. Lubensky and J-H. Chen, *Phys. Rev. B* **17** (1978) 366.

[24] P.G.de Gennes and J.Prost, *The Physics of Liquid Crystals*, 2nd ed. (Clarendon Press, Oxford, 1993).

[25] S. C. Zhang, T. H. Hansson and S. Kivelson, *Phys. Rev. Lett.* **63**, 903 (1989).

[26] A. M. J. Schakel, *Nucl. Phys. B* **453** [FS], 705 (1995).

[27] X. G. Wen, Y. S. Wu, *Phys. Rev. Lett.* **70**, 1501 (1993).

[28] L. Pryadko and S. C. Zhang, *Phys. Rev. Lett.* **73**, 3282 (1994).

[29] E. Babaev, A. Sudbo, and N. W. Aschroft, *Nature* **431**, 666 (2004).

[30] R. Folk, D. V. Shopova and D. I. Uzunov, *Phys. Lett. A* **281**, 197 (2001).

[31] D. V. Shopova, T. P. Todorov, T. E. Tsvetkov and D. I. Uzunov, *Mod. Phys. Lett. B* **16**, 829 (2002).

[32] D. V. Shopova, T. P. Todorov, D. I. Uzunov, *Mod. Phys. Lett. B* **17**, 141 (2003).

[33] D. V. Shopova, T. P. Todorov, *J. Phys. Condensed Matter* **15**, 5783 (2003).

[34] D. V. Shopova, T. P. Todorov, *Phys. Lett. A* **314**, 250 (2003).

[35] , D. V. Shopova, T. P. Todorov, *J. Phys. Studies* **7**, 330 (2003).

[36] T. P. Todorov, *PhD Thesis* (Bulg. Acad. of Sciences, Sofia, 2003).

[37] R. Folk, Yu. Holovatch, in: *Correlations, Coherence, and Order*, ed. by D. V. Shopova and D. I. Uzunov (Kluwver Academic/Plenum Publishers, New York, 1999), p. 83.

[38] I. D. Lawrie, *J. Phys. C: Solid State Phys.* **16**, 3513 (1983).

[39] J. C. Rahola, *J. Phys. Studies*, **5**, 304 (2001).

[40] S. W. Lovesey, *Z. Physik B - Cond. Matter* **40**, 117 (1980).

[41] M. Suzuki and D. I. Uzunov, *Physica A* **204**, 702 (1994).

[42] M. Suzuki and D. I. Uzunov, *Physica A* **216**, 489 (1995).

[43] L. Craco, L. De Cesare, I. Rabuffo, I. P. Takov, D. I. Uzunov, *Physica A* **270**, 486 (1999).

[44] O. Madelung (ed.), *Numerical Data and Functional Relationships in Science and Technology*, New Series, 21, *Superconductors* (Springer, Berlin, 1990).

[45] I. D. Lawrie, *J. Phys. C: Solid State Phys.* **16**, 3527 (1983).

[46] R. P. Huebener, *Magnetic Flux Structures in Superconductors* (Springer, Berlin, 2001).

[47] L. M. Bettencourt and G. J. Stephens, *Phys. Rev. B* **67**, 066105 (2003).

[48] L. M. Abreu, A. P. C. Malbouisson, *Phys. Rev. A* **322**, 111 (2004).

[49] D. V. Shopova, D. I. Uzunov, *Phys. Lett. A* **328**, 232 (2004).

[50] L. M. Abreu, A. P. C. Malbouisson, *Phys. Lett. A* **332**, 153 (2004) (Corrigendum).

[51] L. M. Abreu, C. de Calan, A. P. C. Malbouisson, *arXiv: cond-mat*/0402053.

[52] L. M. Abreu, A. P. C. Malbouisson, I. Roditi, *Eur. Phys. J.* **37**, 515 (2004).

[53] L. Marotta, M. Camarda, G. G. N. Angilella, and F. Siringo, *Phys. Rev. B* **73**, 104517 (2006).

[54] D. I. Uzunov, *Phys. Lett. A* **87**, 11 (1981).

[55] D. I. Uzunov, in: *Advances in Theoretical Physics*, ed. by E. Caianiello (World Scientific, Singapore, 1990), p. 96; *Riken Rev.* **2**, 9 (1993).

[56] I. D. Lawrie, *Nucl. Phys. B* **200** [FS4], 1 (1982).

[57] I. D. Lawrie, *J. Phys. C: Solid State Phys.* **15**, L879 (1982).

[58] C. Dasgupta, B. I. Halperin, *Phys. Rev. Lett.* **47**, 1556 (1981).

[59] J. Bartholomew, *Phys. Rev. B* **28**, 5378 (1983).

[60] H. Kleinert, Lett. *Al Nuovo Cimento* **35**, 405 (1982).

[61] M. E. Peshkin, *Ann. Phys. (N.Y.)* **113**, 122 (1978).

[62] M. Kiometzis, H. Kleinert, and A. M. J. Schakel, *Phys. Rev. Lett.* **73**, 1975 (1994).

[63] M. Kiometzis, H. Kleinert, and A. M. J. Schakel, *Forschr. Phys.* **43**, 697 (1995).

[64] I. F. Herbut, *J. Phys. A: Math. Gen.* **30**, 423 (1997).

[65] C. de Calan and F. S. Nogueira, *Phys. Rev. B* **60**, 4255 (1999).

[66] P. Olsson, S. Teitel, *Phys. Rev. Lett.* **80**, 1964 (1998).

[67] S. Kolnberger, R. Folk, *Phys. Rev. B* **41**, 4083 (1990).

[68] R. Folk, Yu. Holovatch, *J. Phys. A: Math. Gen.* **29**, 3409 (1996).

[69] I. D. Lawrie and C. Athorne, *J. Phys. A: Math. Gen.* **16**, L587 (1983).

[70] I. D. Lawrie, *Phys. Rev. Lett.* **78**, 979 (1997).

[71] F. Freire and D. F. Litim, *Phys. Rev. D* **64**, 045014 (2001).

[72] B. Bergerhoff, F. Freire, D. F. Litim, S. Lola, and C. Wetterich, *Phys. Rev. B* **53**, 5437 (1996).

[73] I. F. Herbut and Z. Tesanovic, *Phys. Rev. Lett.* **76**, 4588 (1996).

[74] J. A. Hertz, *Phys. Rev. B* **18**, 4875 (1978).

[75] J. A. Hertz, *Phys. Scripta T* **10**, 1 (1985).

[76] D. Boyanovsky, J. K. Cardy, *Phys. Rev. B* **25**, 7058 (1982).

[77] D. I. Uzunov, E. R. Korutcheva, and Y. T. Millev, *J. Phys. A: Math. Gen.*, **16**, 247 (1983); 17, 247 (1984); Err., ibid, 17, 1775 (1984).

[78] I. D. Lawrie, Y. T. Millev, and D. I. Uzunov, *J. Phys. A: Math. Gen.* **20**, 1599 (1987); (Err.) ibid, 20, 6159 (1987).

[79] C. Athorne and I. D. Lawrie, *Nucl. Phys. B* **257** [FS14], 577 (1985).

[80] E. R. Korutcheva, Y. T. Millev, *J. Phys. A: Math. Gen.*, **17**, L511 (1984).

[81] G. Busiello, L. De Cesare, and D. I. Uzunov, *Phys. Rev. B* **34**, 4932 (1986).

[82] D. I. Uzunov, E. R. Korutcheva, and Y. T. Millev, *Physica A* **129**, 535 (1985).

[83] D. I. Uzunov, E. R. Korutcheva, and Y. T. Millev, *Phys. St. Solidi* (b) **130**, 243 (1985).

[84] D. Davidov, C. R. Safinya, M. Kaplan, S. S. Dana, R. Schaetzing, R. J. Birgeneau, and J. D. Litster, *Phys. Rev. B* **19**, 1657 (1979).

[85] J. D. Litster, R. J. Birgeneau, M. Kaplan, and C. R. Safinya, in: *Order in Strongly Fluctuating Condensed Matter Systems*, ed. by T. Riste (Plenum, New York, 1980).

[86] C. C. Huang, S. C. Lien, *Phys. Rev. Lett.* **47**, 1917 (1981).

[87] C. W. Garland, G. Nounesis, *Phys. Rev. E* **49**, 2964 (1994).

[88] I. Lelidis, G. Durand, *Phys. Rev. Lett.* **73**, 672 (1994).

[89] M. A. Anisimov, V. P. Voronov, E. E. Gorodetskii, V. E. Podneks, F. Kholmudorov, *Pis'ma Zh. Eksp. Teor. Fiz.* **45**, 336 (1987) [*JETP Lett.* **45**, 425 (1987)].

[90] M. A. Anisimov, P. E. Cladis, E. E. Gorodetskii, D. A. Huse, V. E. Podneks, V. G. Taratuta, W. van Saarloos, and V. P. Voronov, *Phys. Rev. A* **41**, 6749 (1990).

[91] A. Yethiraj and J. Bechhoefer, *Phys. Rev. Lett.* **84**, 3642 (2000).

[92] J. A. Hertz, *Phys. Rev. B* **14**, 1165 (1976).

[93] D. V. Shopova, D. I. Uzunov, *Phys. Rep. C* **379**, 1 (2003).

[94] M. K. Bushev, D. I. Uzunov, *Phys. Lett. A* **76**, 306 (1980); Err., ibid, 78, 491 (1980).

[95] D. I. Uzunov, *Phys. Lett. A* **78**, 395 (1980).

[96] P. C. Hohenberg, B. I. Halperin, *Rev. Mod. Phys.* **49**, 435 (1977).

[97] M. P. A. Fisher, G. Grinstein, *Phys. Rev. Lett.* **60**, 208 (1988).

[98] M. P. A. Fisher, G. Grinstein, and S. M. Girvin, *Phys. Rev. Lett.* **64**, 587 (1990).

[99] I. F. Herbut, *Phys. Rev. B* **57**, 13729 (1998).

[100] D. Matthey, N. Reyren, and J.-M. Triscone, and T. Schneider, *arXiv:cond-mat*/0603079.

[101] T. Schneider, *arXiv:cond-mat*/0610230030.

[102] M. A. Continentino, *Quantum Scaling in Many-Body Systems* (World Scientific, Singapore, 2001).

[103] N. S. Tonchev, D. I. Uzunov, *J. Phys. A: Math. Gen.* **14**, 521 (1981).

[104] N. S. Tonchev, D. I. Uzunov, *J. Phys. A: Math. Gen.* **14**, L103 (1981).

[105] N. Grewe and B. Schuh, *Z. Phys. B* **36**, 89 (1979).

[106] N. Grewe and B. Schuh, *Phys. Rev. B* **22**, 3183 (1980).

[107] P. Bhattacharyya, *J. Phys. C: Solid State Phys.* **16**, L1011 (1983).

[108] L. Radzihovsky, A. M. Ettouhami, K. Saunders, and J. Toner, *Phys. Rev. Lett.* **87**, 027001 (2001).

[109] A. M. Ettouhami, K. Saunders, L. Radzihovsky, and J. Toner, *Phys. Rev. B* **71**, 224506 (2005).

[110] E. J. Blagoeva, G. Busiello, L. De Cesare, Y. T. Millev, I. Rabuffo, and D. I. Uzunov, *Phys. Rev. B* **42**, 287 (1990).

[111] Y. T. Millev, D. I. Uzunov, *Phys. Lett. A* **145**, 287 (1990).

[112] G. Busiello, L. De Cesare, Y. T. Millev, I. Rabuffo, and D. I. Uzunov, *Phys. Rev. B* **43**, 1150 (1991).

[113] D. V. Shopova and D. I. Uzunov, *Phys. Rev. B* **72**, 024531 (2005).

[114] D. I. Uzunov, *Phys. Rev. B* **74**, 134514 (2006).

[115] D. I. Uzunov, *Europhys. Lett.* (2006); in press.

[116] D. I. Uzunov, E17-2006-128 (*Preprint of JINR-Dubna*).

[117] E. Brèzin, D. R. Nelson, and A. Thiaville, *Phys. Rev. B* **11** (1985) 7124.

[118] I. Affleck, E. Brésin, *Nucl. Phys. B* **257** [FS14], 451 (1985).

[119] L. Radzihovsky, *Phys. Rev. Lett.* **74**, 4722 (1995).

[120] E. Brézin, A. Fijita, and S. Hikami, *Phys. Rev. Lett.* **65**, 1949 (1990).

[121] G. P. Mikitik, *Zh. Eksp. Teor. Fiz.* **101**, 1042 (1992) [*Sov. Phys. JETP* **74**, 558 (1992)].

[122] L. Radzihovsky, *Phys. Rev. Lett.* **76**, 4451 (1996).

[123] I. F. Herbut and Z. Tesanovic, *Phys. Rev. Lett.* **76**, 4450 (1996).

In: Progress in Superconductivity Research
Editor: Oliver A. Chang, pp. 55-78

ISBN: 1-60021-694-3

Chapter 2

EEM: The Exciton Enhancement Mechanism Theory and Experimental Evidence of Optically Enhanced Tc in High Tc Superconductors

K.W. Wong [*] ***and S. Curatolo*** [**]
CZT Inc. Lawrence Kansas 66049 USA

Abstract

The Excitonic Enhancement Mechanism [1] is proposed as the theory for superconductivity in some novel hole carrier systems, such as M_3C_{60} [2], cuprate high temperature superconductors [3] as well as the most recently discovered Boron doped Silicon [4]. It is suggested that because excitons are like hydrogen atoms and possess excited energy levels, these Cooper pair exchange particles can be excited by suitable infrared light [5], resulting in large enhancement of the superconductivity transition temperature [5]. Experimental data on the EEM Tc enhancement was shown for bulk and thin film YBCO. With the optimum YBCO bulk system (inside an open liquid Nitrogen cryogenic chamber), we show the enhancement by observing the diamagnetism before and after IR light was applied. The transition temperature was shown to increase ~80°K, from $93^{o}K\ to\ \geq\ 175^{o}K$ exactly as predicted by the EEM theory [5]. The results of these original measurements were verified on an oxygen deficient YBCO thin film by measuring the transition temperature using AC voltage versus temperature measurement (inside a closed Helium cryogenic chamber) carefully obtained by using a 100Khz DSP lock-in-amplifier, before and after IR light was applied. The transition temperature was shown to increase by ~17°K, from 69.5°K to 86.5°K[5]. The increase in Tc, although small, is shown to be exactly as the value calculated by the same EEM theory [5]. It is therefore without a doubt that these results confirm EEM as the theory for cuprates and its application to high Tc superconductors.

[*] Kai Wai Wong, Professor Emeritus University of Kansas
[**] Corresponding author: Susana Curatolo, President CZT Inc. czt@sunflower.com

1. Introduction

Since the discovery of High Temperature Cuprate Superconductors [6], there has been an expanded effort to find other novel superconducting systems, which include M_3C_{60} [7] and the recently discovered Boron doped cubic Silicon [8]. Most people consider M_3C_{60} and Si-doped B systems' mechanism to be BCS. We believe they are EEM-Excitonic Enhancement Mechanism, as their charge carriers are predominantly p-state holes, and both have an empty s-state conduction band above the valence p-band. Their relatively low Tc values, according to EEM, were a result of smaller band gap and excitonic binding energies. These two new systems have some common features with the Cuprates, such as their charge carriers are mainly 'p' orbital valence band holes, and that their conduction band is an empty 's' orbital state. It is commonly believed though, that, M_3C_{60} and Boron doped Silicon superconducting mechanisms are the electronic structure independent BCS [9] [1], while no consensus has been established for the cuprates. It is worthy to note that the main reasons pointed to their differences are first, the substantially higher Tc value for some cuprates, and secondly, the presence of the 'd' orbitals in cuprates, which is absent in the other systems. In the cuprates, it has been found also that the superconducting Tc value is correlated to the number of CuO_2 layers in its crystal structure, with a maximum reached at 3 layers. On top of this feature Tc varies according with hole carrier density in an inverse parabolic form. The Excitonic Enhancement Mechanism or EEM [1] proposed for the cuprates, also possesses these features.

The basic EEM is based on a two band model, with a partially filled valence 'p' band, and an empty 's' conduction band, separated by a positive band gap G. In fact this theory model is satisfied by the actual band structures of all the above-mentioned novel superconductors. Apart from Tc, the EEM model also predicts other novel properties in both the normal and superconducting phases, such as the sign changes that were observed in the mixed-state quantum Hall effect [10], and the thermal electric power [11]. As an in-depth analysis of the cuprates, a more detailed EEM theory was developed [1] based on a specific CuO_2 plane, separated from the Ba/Sr plane along the c-axis, with the Ba/Sr at the center of the CuO_2 cell [12]. It has been found that this specific geometric model of EEM was able to produce the following properties known to be associated with cuprates: (1) Tc correlation to carrier density and the number of CuO_2 layers [1], as mentioned earlier; (2) the two sign carriers like behavior for both thermal electric power [11] and mixed state quantum hall effect [13]; (3) the existence of perpendicular nodes in the superconducting energy gap [14] in YBCO epitaxial film. Despite these successes of the model there remains no direct proof of the existence of excitons in these systems. It is, therefore, our intention in this paper to show the presence of excitons. The exciton contribution to superconductivity can be verified indirectly through infrared light irradiation. In section 2 of this work, we will present the theoretical analysis, followed by sections 3 through 5 where we present the experimental results supporting the theoretical calculations based on results from infrared irradiation on YBCO systems.

2. Theoretical Analysis of the EEM Theory

For the detailed discussion of the EEM Theory [1], particularly the two dimensional model [1] developed specifically for the cuprate systems, we refer our readers to Wong and Ching [1]. Since the aim of the current paper is to verify this theory's validity for the YBCO system we will discuss our complete analysis based on this specific model of the EEM Theory only. This model is transportable across High Tc Superconducting Cuprates, such as BiSCCO [5]. For analysis of other possible systems, such as M_3C_{60} or Boron doped Silicon, a different geometrical model of EEM must be developed first.

In order to investigate further if EEM can predict new properties in cuprates not yet discovered, let us review the theory. The EEM theory is based essentially on the dipolar interband interaction between an intrinsic partially filled valence 'p' band from the Oxygen ions and the empty conduction 's ' band of the Ba/Sr ions, which lay above the CuO_2 layer. Because of this interband interaction, a significant renormalization of the intrinsic electronic band structure is produced, and excitonic levels are introduced inside the band gap. In fact excitonic levels within the band gap occurs in Cu_2O crystal as well [15]. In the HTS cuprates, according to EEM the excitonic ground level lies just above the valence band VB, while the renormalized hole band is strongly space dependent [1]. Due to this feature, it was found that much of the occupied 'p' electrons in VB are located along the a,b axes of the unit cell, separated from the excitonic ground level by G* inside the band gap G. As G* is of order 100°K, these 'p' electrons are thermally excitable into this G* level. Should this happens, global localized excitons having binding energy exceeding thermal fluctuation are formed. These Bosonic excitons can now be employed as exchanged particles by the mobile holes to form Cooper pairs, similar to the phonons mechanism in the Low Temperature Superconductors. Since excitons have excited levels similar to phonons, we therefore expect that if a large enough population of such excited excitons are maintained, Tc can be enhanced [5] [16]. We refer our readers to reference [1] for in-depth discussion of the background.

For a more detail analysis, let us refer back to the renormalized band structure results as given in reference [1]. The renormalized valence band hole energy spectrum λ_2 is given by

$$\lambda_2 = \varepsilon_F^* - \varepsilon_a - \Delta_{ps} \tag{1}$$

where ε_a is the hole kinetic energy, and ε_F^* is the renormalized Fermi-energy, which cuts through VB.

$$\varepsilon_F^* = \varepsilon_F - \frac{\Delta_{dip}^2}{\varepsilon_F + G^*} \tag{2}$$

Δ_{dip} is the space dependent dipolar transition matrix. [See equation (3.35) in reference [1], for detail], and G* the renormalized band gap

$$G^* = G - E_o \tag{3}$$

The quantity E_o is the ground state of the TWO-dimensional exciton binding energy,

$$E_o = \frac{16Rm^*}{\varepsilon^2} n^2 \tag{4}$$

where R is the Rydberg constant, ε is the interband dielectric constant, m* the effective reduced mass and 'n' the fraction of 'p' hole density. Obviously, the 2d-exciton has excited energy levels within the renormalized band gap. These levels are given by

$$G_m^* = G - E_m \tag{5}$$

where

$$E_m = \frac{E_o}{(2m+1)^2} \quad , \text{ m=1,2,...} \tag{6}$$

These higher levels are of order eV above VB, and are not reachable by thermal excitation. Finally the pseudo-gap Δ_{ps} [see equation (3.40) in reference [1]] is positive, but vanishes on ε_F^*. Now let us turn to the space dependent form for Δ_{dip} [see equation (3.37) in reference [1]]

$$\Delta_{dip}^2 = \Delta^{(o)2} [1 - \sin 2\phi] \tag{7}$$

where ϕ is the angle measured from the a-axis of the cell in the a-b plane. It is clear that the renormalized hole Fermi energy ε_F^* as shown in equation (2), varies from a maximum of ε_F to a minimum of $\varepsilon_F - \dfrac{\Delta^{(o)2}}{\varepsilon_F + G^*}$ along the a,b axes. In another word the hole density is largest along the cell diagonals, and smallest along the a,b axes[25], resembling a liquid with surface tension being placed in a rectangular container, with the container walls given by the ac, and bc surfaces. Such a fluid will have a fluid surface lowest at the center of the ab cross-section, and highest on the walls. The difference between the lowest and highest surface points is determined by the surface tension. If the height of the container walls as measured from the lowest fluid surface level is less than the necessary difference due to the surface tension, then fluid will creep over the walls, until a new equilibrium level is reached. For the 'p' electrons in VB, it is therefore those along the a,b axes that are first excited into the excitonic ground level G*, provided the system temperature T is of order G*. However, the thermal excitation of the electrons into G*, does not mean that stable excitons are formed. The excitons are stable if and only if, its binding energy exceeds the thermal fluctuation energy, which creates them in the first place. Should this condition be satisfied, global formation of localized excitons will occur in the structure, quite similar to the phonon lattice. Because excitons are Bosons, when coupled to the mobile holes can replace the phonons in the BCS theory, and lead to hole-hole Cooper pairing, thus producing superconductivity.

To obtain Tc based on exchange of excitons, we first obtain the exciton's excitation energy, [see equation (3.15) in reference [1]], which is given by λ_3 minus the exciton binding energy to the lattice

$$\hbar\Omega = \lambda_3 - \frac{2\Delta_{dip}^2}{\varepsilon_F + G^*} \tag{8}$$

where

$$\lambda_3 = G^* + \varepsilon_F^* + \frac{2\Delta_{dip}^2}{\varepsilon_F + G^*} \tag{9}$$

[Note, the excitonic binding to the lattice vanishes along the cell diagonals. The space average value of which is of order eV, and is quite stable in the superconducting phase. Hence the lateral thermal fluctuation that destabilizes the bound exciton occurs in the normal phase. Because the excitonic levels of the system are inside the band gap, it implies such exciton is localized or bounded to the crystal lattice. Hence, when the exciton is delocalized, it will decompose into a free electron and a free hole. Such a picture for the normal phase gives raise to the two carriers model [17].]

This bound excitonic excitation energy $\hbar\Omega$ now replaces the Debye phonon energy in the Cooper pair BCS energy gap

$$\Delta_{BCS} = 2\text{h}\Omega \exp-\left(\frac{1}{JN_F^*}\right) \tag{10}$$

where J is the coupling strength and N_F^* is the hole Fermi-surface density.

Since the superconducting Tc is directly proportional to the BCS gap Δ_{BCS}, it follows for the cuprate, Tc is proportional to the exciton excitation energy. It is precisely this feature that allows EEM to explain the Tc dependence on the hole density [see reference [1] for more detail]. Suppose we excite the excitons to its first excitation level optically, then it follows that Δ_{BCS} will also be enhanced, as $\hbar\Omega$ is replaced by

$$\hbar\Omega_1 = G^*(1) + \varepsilon_F^*(1) \tag{11}$$

Before going further, let us discuss the photo-excitation necessary to excite the excitons into the first excited level, we need a photon of energy [5]

$$\hbar\gamma_1 = \frac{8}{9}E_o \tag{12}$$

This photon energy is the same for exciting the system from either above or below Tc, as $G^* \cong 0$. Since $\hbar\gamma_1$ is of order 1 eV, it follows the photon wavelength is much larger than the crystal cell dimensions. Hence when we pump the system with such IR photons, the populated electrons with the highest energy level are excited. There is a difference where these electrons are located depending on the system's physical phase. Suppose the temperature T of the system is below Tc, which means stable ground state excitons are already formed, then it follows that the photon must directly excite the excitons into the level G*(1).

$$G^*(1) = G - \frac{E_o}{9} \tag{13}$$

which in turn changes the renormalized Fermi energy to

$$\varepsilon_F^*(1) = \varepsilon_F - \frac{\Delta_{dip}^2}{\varepsilon_F + G^*(1)} \tag{14}$$

Following the discussion earlier on the proportionality of Tc on the exchange exciton excitation energy $\hbar\Omega_1$ and assuming J and N_F^* remain relatively constant, we can estimate the enhancement of Tc to Tc* from the ratio

$$\frac{T_c^*}{T_c} \cong \frac{\hbar\Omega_1}{\hbar\Omega} \tag{15}$$

How much Tc is enhanced to Tc* depends critically on the sample's original Tc value, because Tc of the sample is dependent on the 'p' hole density n as given in equation [3.63] of reference [1], through the excitonic excitation energy

$$\hbar\Omega = -\frac{16Rm^*}{\varepsilon^2}(n - n_o)^2 + \overline{E_o} + G - \frac{\left\langle \Delta_{dip}^2 \right\rangle_\phi}{\dfrac{-16Rm^*}{\varepsilon^2}(n - n_o)^2 + \overline{E_o} + G} \tag{16}$$

where $\overline{E_o}$ is the optimum excitonic ground state binding energy [17] [5]

$$\overline{E_o} \equiv \frac{16Rm^*}{\varepsilon^2} n_o^2 \tag{17}$$

In YBCO, we found [5]

$$\left\langle \Delta_{dip}^2 \right\rangle_\phi = \frac{\pi - 2}{\pi} \Delta^{(o)2} \tag{18}$$

where $$\Delta^{(o)} = 2.07\ eV \cong 2\overline{E_o} \tag{19}$$

To facilitate our study of infrared light enhancement on the critical temperature, namely Tc*, we will consider two separate cases. First, the optimum YBCO sample, when $n = n_o$, and Tc is maximum, which corresponds to $93^\circ K$.

$$\hbar\Omega_o = \frac{4}{\pi}\overline{E_o}$$

as $$E_o = \overline{E_o} \cong G \cong \varepsilon_F/2 \tag{20}$$

By substituting these into equation (11), we obtain

$$\hbar\Omega_1(n_o) = \left(\frac{26}{9} - \frac{\pi - 2}{\pi}\frac{18}{13}\right)\overline{E_o} \tag{21}$$

Hence from equation (15), we get

$$Tc^*(n_o) = 93 \times \frac{\left(\frac{26}{9} - \frac{\pi-2}{\pi}\frac{18}{13}\right)}{4/\pi} = 174^\circ K \tag{22}$$

For the $Tc = 70^\circ K$, which is oxygen deficient, we need to find the oxygen deficient factor $\Delta\eta \equiv (n_o - n)/n_o$ for this sample first. By substituting it into equation (16), $\Delta\eta$ and comparing $T_c(n)$ *to* $T_c(n_o)$, we get

$$\frac{T_c(n_o) - T_c(n)}{T_c(n_o)} = \frac{\pi}{4}(\Delta\eta)^2 + \frac{\pi - 2}{\pi}\frac{(\Delta\eta)^2}{2\left(2 - (\Delta\eta)^2\right)} \tag{23}$$

leading to $\Delta\eta = 0.53$

Substituting $\Delta\eta = 0.53$ into E_o, and into equation (11), we obtain

$$\hbar\Omega_1(n) = -\overline{E_o}(\Delta\eta)^2 + \frac{8}{9}\overline{E_o}(1 - \Delta\eta)^2 + 2\overline{E_o} - \frac{\pi - 2}{\pi}\frac{4\overline{E_o}}{2 - (\Delta\eta)^2 + \frac{8}{9}(1 - \Delta\eta)^2} \tag{24}$$

By comparing $\hbar\Omega_1(n)$ to $\hbar\Omega_o(n_o)$, we have

$$\frac{T_c*(n)}{T_c(n_o)} = \frac{\left\{2-(\Delta\eta)^2+\frac{8}{9}(1-\Delta\eta)^2-\frac{\pi-2}{\pi}\frac{4}{2-(\Delta\eta)^2+\frac{8}{9}(1-\Delta\eta)^2}\right\}}{4/\pi} \tag{25}$$

Thus, with $\Delta\eta = 0.53$, we obtain

$$T_c*(n) = 86^{\circ}K$$

This result shows explicitly how much T_c* is reduced in oxygen deficient samples. In fact in this particular case, the infrared light is unable to enhance T_c very much, not even to the $T_c(n_o)$ value, thus leading to a possible misconception that infrared irradiation cannot enhance the superconducting transition temperature in non-optimum samples.

3. The Effects of Infrared Irradiation

Before the discussion on the Tc* experimental measurements which we will present in the next section, we need to fully study the effects of the proper infrared irradiation on the cuprate superconductor samples.

Because excitons contain hydrogen like excitation spectra, it is excitable by photons of the proper wavelength. The exciton energy levels are given by

$$E_m = -\frac{E_o}{(2m+1)^2} \qquad m = 0,1,2,... \tag{26}$$

as measured from the bottom of the 's' conduction band. Therefore the photon energy needed to excite a 'p' electron from the a-b axes in the valence band, such that the mth level excited exciton can be reached is

$$\hbar\nu_m = G - \frac{E_o}{(2m+1)^2} \tag{27}$$

In a hydrogen atom the energy levels are sharp, while in a solid the exciton energy levels are broadened by the solid's structure symmetry, thus we can use a range of photon wavelengths to create excited excitons, namely

$$G > \frac{hc}{\lambda} > G - \frac{E_o}{9} \tag{28}$$

As $E_o = \overline{E_o}(1 - \Delta\eta)^2$, where $\Delta\eta$ is the oxygen deficiency factor, and $\overline{E_o} \cong G$, it follows

$$\frac{\frac{9}{8}\frac{hc}{G}}{\left\{1 + \frac{1}{4}\Delta\eta - \frac{1}{8}\Delta\eta^2\right\}} > \lambda > \frac{hc}{G} \tag{29}$$

It is clear that, the proper λ must lie within these limits. For an optimum sample, the λ range is largest, while for hole deficient samples the λ range becomes very narrow as $\Delta\eta$ exceeds 0.5. This condition is reflected in the small Tc enhancement of such samples. Clearly the infrared wavelength λ cannot exceed either limits to produce the Tc enhancement. Infrared irradiation with λ below hc/G will destroy the excitons, thus, superconductivity; while λ above $\frac{9}{8}hc/G$ is too weak to create excited excitons, and thus, unable to enhance Tc. Obviously λ is not the only requirement. We also need to understand how the photons are absorbed and the necessary intensity of the irradiation. There are differences between surface absorption and interior absorption in the sample. For example, for a thin film sample, the predominant absorption is surface absorption, with minimal secondary absorption due to backscattering from below the surface. While in a bulk polycrystal sample, much of the photon absorption occur inside the sample due to backscattering and inter-grain scattering. Thus, we believe the absorption rate is higher in bulk polycrystal than in thin film. Let us now turn to the question of excited exciton population.

Normally this excited excitonic state bound to the lattice is unstable, but we can maintain the population through continuous optical pumping. Thus, let us investigate the equilibrium excited exciton population through optical pumping. Suppose N(o) is the maximum allowed number density of excitons in the system. Let N(t) be the number density of excited excitons at time t. Then

$$\frac{dN}{dt} = \alpha_I\left(N(o) - N\right) - \beta N \tag{30}$$

where α_I is the optical intensity dependent excitation rate, and hence varies from the surface to the inside of the sample. β is the natural relaxation rate of the excited state. We obtain

$$N(t) = \frac{\alpha_I}{\alpha_I + \beta} N(o)\left[1 - e^{-(\alpha_I + \beta)t}\right] \tag{31}$$

Since α_I increases with the light intensity, while β is the intensity independent relaxation rate, the sharpness of the superconducting transition Tc* must then be adjustable through increasing the IR beam intensity. The equilibrium excited exciton density is thus

$$N(\infty)=\frac{N(o)}{1+\beta/\alpha_I} \tag{32}$$

The ratio β/α_I must be less than a certain critical value to provide sufficient number of excitons to facilitate Cooper pairing. As α_I changes with depth from the sample surface, it is possible for us to observe diamagnetism at an elevated temperature above Tc, yet cannot obtain zero resistivity and vice versa, as these two phenomena come from different parts of the sample. Electron conduction is purely surface, while dia-magnetism is a bulk property. None-the-less, by increasing the photon intensity we still expect to be able to observe both phenomena simultaneously. Photo-absorption by any substance creates heating, while exciton relaxation by photo-emission cools the sample. Thus, we expect the sample surface temperature may be lower or higher than its body temperature. In measuring Tc* using resistivity measurement it is important to measure the surface temperature. This value can be tens of degrees lower or higher than the bulk temperature, depending on the cooling or warming temperature rate, which is in equilibrium with the thermal bath. With these discussions we now turn to the experimental part of this paper.

4. Experimental Setup & Data for Thin Film sample Geneva III Experiment

The experiment was conducted to test the theoretical prediction of the superconducting state of YBCO at an enhanced temperature Tc* above the well known critical temperature Tc for YBCO. In obtaining the Tc* value from resistivity to temperature measurement, the important conditions are: (1) the uniformity of the sample, not the optimum Tc value, as that can be accurately calculated as shown in section 2 of this paper. Uniformity of the sample is shown by the sharpness in the phase transition at Tc. (2) Uniformity of the temperature through out the sample during the R-T measurement. Therefore, a thin film sample is most ideal. For this purpose, readily commercially available HTS thin film YBCO sample[18] was used. The sample was Theva-S type HTS YBCO, of square shape: 8mm x 8 mm x 3000 Angstroms; manufactured by THEVA [19]. The sample was mounted inside the Janis [20] cryogenic chamber/tower. The test was designed to accommodate easy access of the light source to induce IR pumping (corresponding to ~0.9eVor 1300 nm) [5] and for this reason an access quartz window on one side of the cryostat setup could be used for the test. The coolant was liquid helium. The temperature, ac micro-voltage current, and phase were measured by a SR830-100 kHz DSP lock-in amplifier [21]. This SRS Lock-in-amplifier was also used to provide AC micro-voltage and micro-current to the sample inside the cryogenic chamber to an accuracy of about ±1% throughout the experiment temperature range. The experimental procedures were filmed throughout the tests and are available on DVD.

1. Sample setup

Sample was held inside the cryogenic chamber by 4 bolts mounted on a copper plate onto a brass spacer so as to bring the sample closer to the 1.63" diameter x 5 mm thick quartz glass window. It was centered onto 1" diameter inner window and its distance from the outer window inner wall was 22 mm as a result of improvement over the original distance (before using the spacer) of 37 mm. Probes on the sample were made of gold. On the square sample, the outer probe strips are for current and inner probe strips are for voltage. See Figure 4.1 and inset schematic below. Due to the sample size being so small special equipment was required to install properly the probes and secure them to the mounted sample and prevent them from detaching, as seen in Figure 4.1.

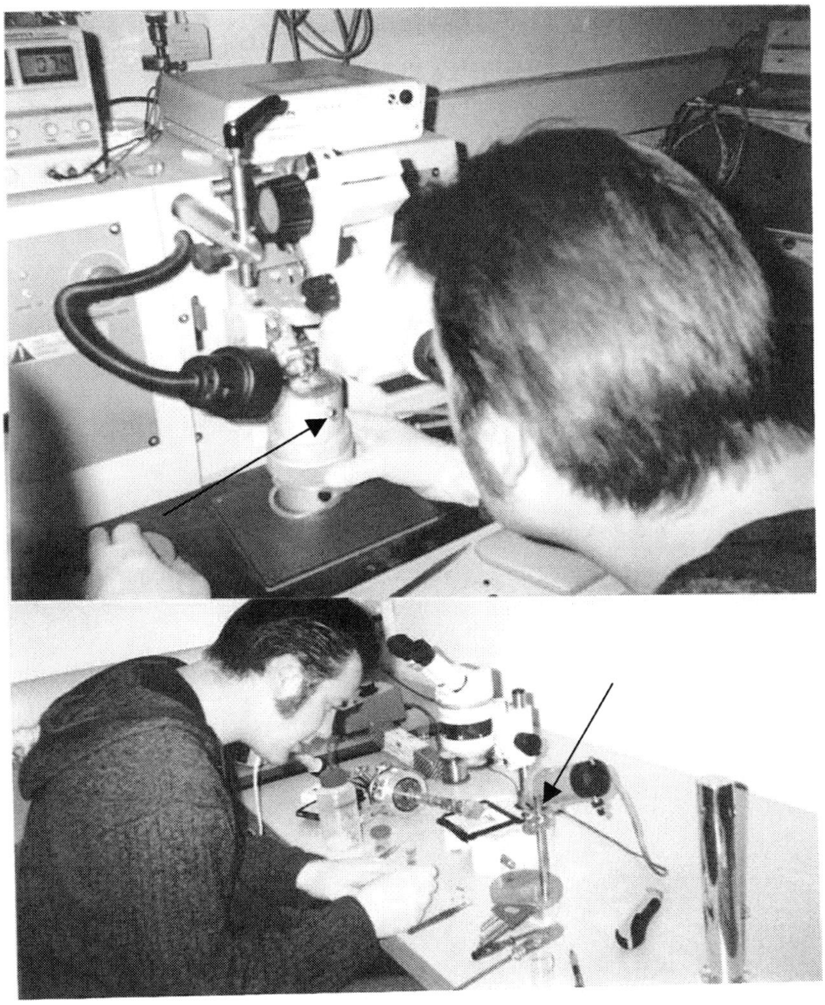

Figure 4.1. Continued on next page.

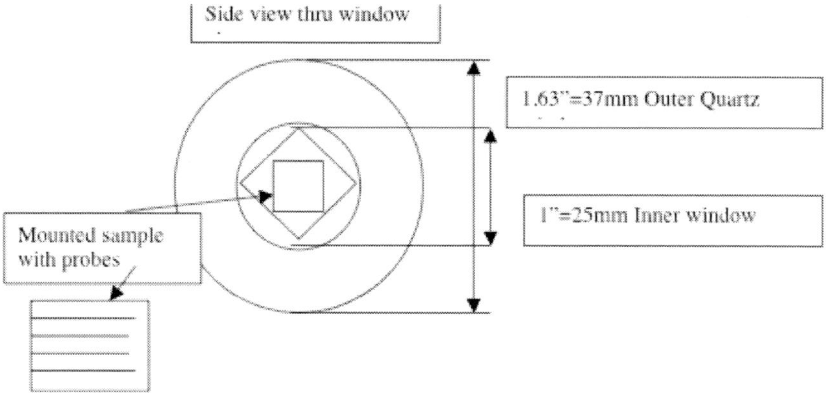

Figure 4.1. Theva sample probe mounting (top) and distance to window inserts adjustment (bottom). Inset: Schematic of Theva sample probe as seen by IR light source [22] [5].

2. Light Source Setup

Figure 4.2. Side view close-up of mounted Theva sample inside the Janis cryogenic chamber with IR source butted against chamber window facing the sample surface. Mounting is off center accommodating the modification of the distance of the sample surface to IR light source. [18] [5]

IR light of wavelength of ~0.9eV or 1300nm; Bundle of 10 LEDs Marubeni type C of 1" (25mm) diameter total- butted against outer wall of 1.63" diameter x 5 mm thick quartz glass window facing the square mounted sample inside cryogenic chamber. Light source was powered by 9 dc volts and mounted on Styrofoam square holder attached to a chemistry arm used to hold and align it. The camera was used to view the bundle before sample inserted to align it from opposite viewing side of the chamber so that the center diodes would line up with the 8x8 mm square sample provided. See figure 4.2:

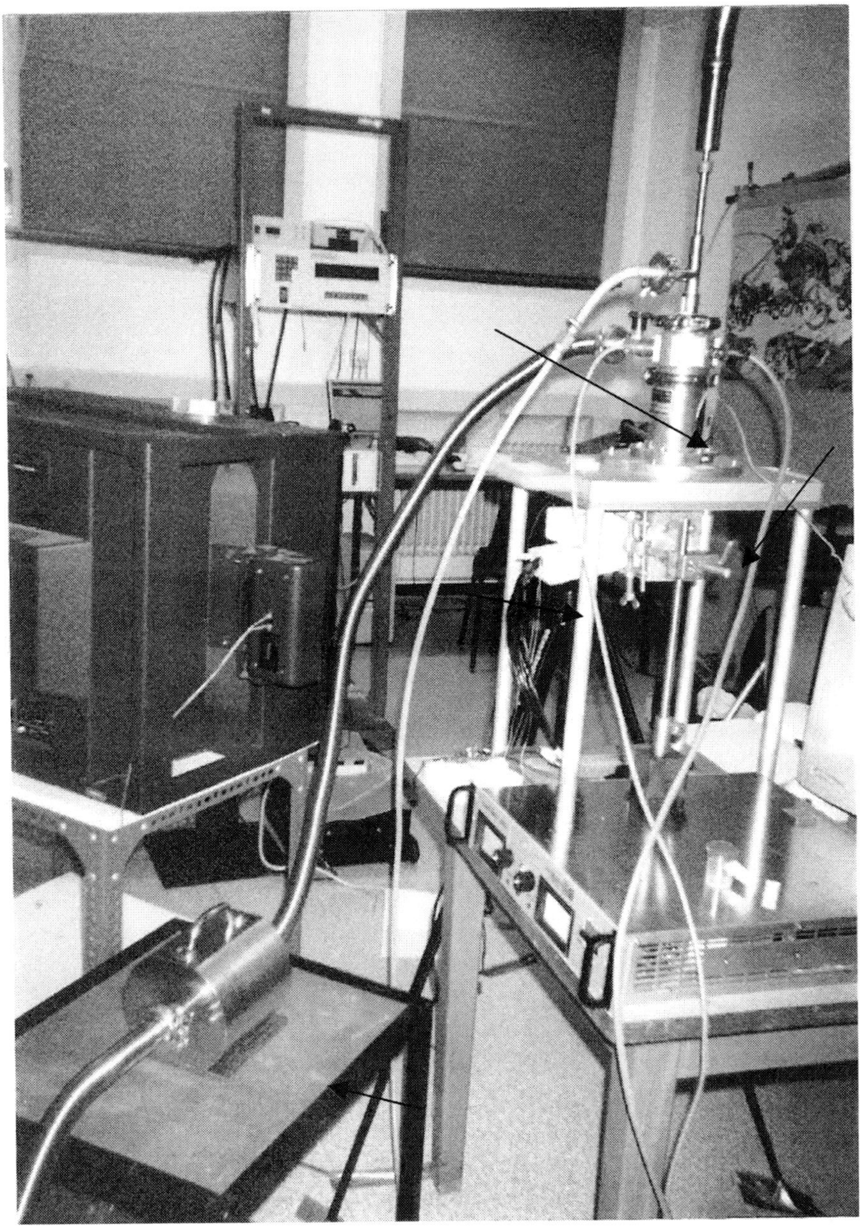

Figure 4.3. Back view of the Janis cryogenic chamber and the helium attachment with IR source butted against chamber window [22] [5].

3. Set-up of SRS Lock-in-Amplifier

- Time constant: 18dB/x100/ms
- Signal input:A.B/AC/ground
- Sensitivity:5/x10/microvolts
- Reserve:low noise
- Filters:line/2 x line
- Temperature:degrees Kelvin
- Phase:variable
- Frequency:277.0 Hertz
- Start temp=294.6°K
- Start volt=51.96 microvolts
- Constant current=10 microamps
- Cool-down rate=9°K/min→4°K/min thru 2 forced 5°K/min cool-downs: 1rst at row=274, 2nd at row=310 of data set.
- Warm-up rate=9°K/min. Helium shut down started at row=570→row=585(helium totally off) of data set.
- EndOfTape at Temp=287.4°K(row=745) of data set.
- Total Run: 745 rows (10seconds = 1 row)=7450 seconds
- Data sampling rate= 10 seconds

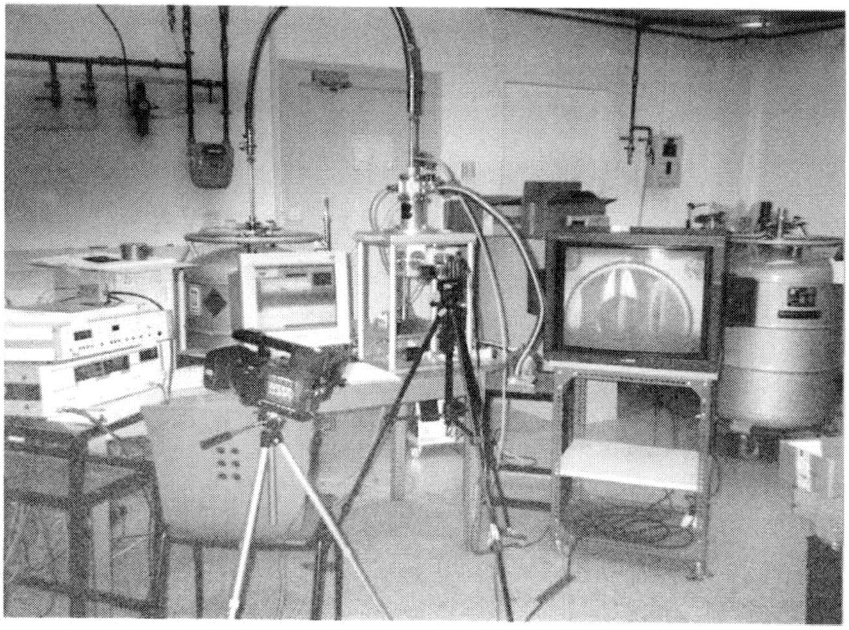

Figure 4.4. Experimental setup for GenevaIII: From left to right: SRS lock in amplifier being filmed by Panasonic CCD camera(large camera); desktop computer interface-data collection from SRS and Janis cryogenic chamber with Hellium attachment, containing Theva-sample and IR source butted against chamber window; Sony video camera(small camera) filming sample alignment and showing it on TV. [18] [5].

1. R-T Measurement in the Superconducting State at T>Tc

The enhanced critical temperature Tc* was tested by using a four point probe to measure the ac microvolts of the HTS YBCO thin film as a function of temperature. The verification of zero resistance in a very small sample is not easy, because YBCO is a relatively good conductor in its normal state and its total resistance is quite small. Thus, many factors can obscure our ac microvoltage measurement. For example, the accuracy of the probe mounting on such a small and thin sample, the presence of Eddie currents, the potential presence of vortex dynamics, etc. The inset schematic of Figure 4.1 and in Figure 4.2, show the alignment, orientation and distance of the 4-point probe on the underside of the Theva sample. The two outer wires of the 4-point probe are designed to apply current, the inner wires are to measure the voltage drop. The current applied was 10 micro-amperes by making use of a ac-power supply operating in the current limited mode built-in the SRS, as well as the micro-voltmeter, phase meter and thermocouple reader with a resolution of 5x10 micro-volts and +/-1% degrees K. The incoherent light source of wavelength 1300 nm, is a hand-made set of Marubeni Infrared Type-C LEDs. The general setup for all the GenevaIII experiment is shown in Figure 4.4.

2. Tc Enhancement by Optical Pumping : GenevaIII Results

The experimental results of Tc enhancement in YBCO by IR light application can be summarized as follows:

Greater detailed information is gained about the onset of the enhanced superconducting state in thin film HTS YBCO when measuring voltage as a function of temperature. As seen clearly in Figures 4.5 the re-introduction of the superconducting state measured as zero voltage occurs at the temperature of 87 °K upon application of IR light. This result was achieved first by cooling the thin film sample to below its Tc value. The oxygen deficient sample has a sharp Tc of 69°K, as shown from the cooling curve in Figure 4.5. When the sample reached ~85°K and 24.4 microvolts at ~ 24 minutes into the cool down at 9°K/min, the voltage was decreasing much too slowly. Therefore, in order to speed up and find the Tc of the Theva sample, we increased the rate of cooling another 5°K/min at the 80°K and 22 microvolts point. We held that steady for ~ 6 minutes and did a second rate of cooling increase of 5°K/min at the 75°K and 18 microvolts point, holding that cooling rate steady all the way to Tc of 69°K and 0 microvolts. During these initial changes of cooling rate, two jerks were introduced thermally, which were expected, and these jerks cannot change the Tc value. This is made clear in Figure 4.5. Upon reaching Tc, we wait a few minutes at which point IR light was applied and the heating of sample began. From 69°K to 87°K the voltage remained zero. At 87°K the system started going out of superconducting phase and reached the complete normal phase at 95°K, and then, increased linearly and parallel to the V-T cooling curve of the sample where no IR light was applied. This constant shift between these two curves shows without a doubt that the carrier density was increased by IR irradiation in the normal phase. If EEM had not happened the cool down and the warm up curves should lie right on top of each other. Thus, this 17°K shift, and the lowering of normal phase resistivity, indicates clearly the presence of EEM. In this experiment the measured and the calculated values for Tc* = 87°K for the thin film sample match (see equation 25, section 2 of this paper).

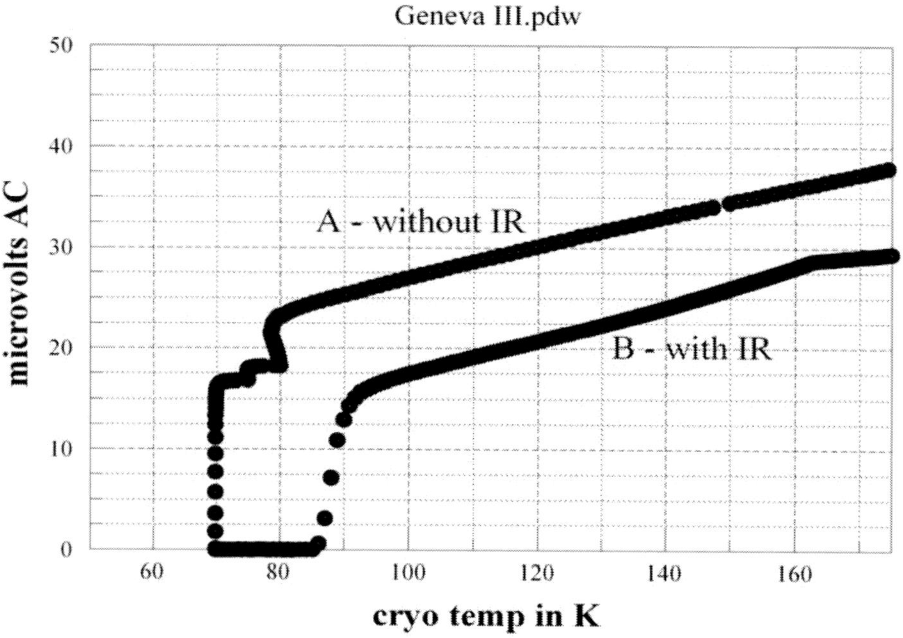

Figure 4.5. GenevaIII V-T data set: Curve A = cool down , Curve B = warm up [5].

5. Experimental Setup and Data for Bulk Sample Kansas III

Experiments

Superconductivity is defined by zero resistivity and diamagnetism. Since it was not possible for us to apply IR irradiation on the sample while it were placed inside an inductance coil for a direct magnetic susceptibility measurement, we were led to using observation of the actual displacement of a magnet place above the superconductor. To do this, we must employ a less accurate open system method. Detail of this experiment is described below.

Two basic experiments were conducted to test the theoretical prediction of the superconducting state of YBCO at an enhanced temperature Tc* above the well known critical temperature Tc for YBCO. For this purpose we used 2 bulk YBCO disks [23], one without a probe for experiment 1, and one with a 4-point probe mounted on its surface for experiment 2. A cube shaped neodymium rare earth magnet was used in the Meissner effect experiment. The disks or tablets had a diameter of 2.45 cm and a thickness of about 3 mm. The tests were designed to accommodate easy access of the light source to induce IR pumping (corresponding to ~0.9eV or 1300nm) [5] and for this reason no enclosed cryostat set up could be used for these tests. The coolant was liquid nitrogen and the temperature was measured by Fluke multi-meter RMS 87V and RMS 189, to an accuracy of about ±1% throughout the temperature range. However, the Fluke thermometer has a lower range limit of operation accuracy of -150°C or 123°K, thus any temperatures below such, including Tc of

93°K, cannot be used with reliability. However, if EEM is valid, the enhanced superconducting state will last all the way up in temperature to over 170°K, way within the accurate range of the Fluke thermometer. The experimental procedures were filmed throughout the tests and are available on DVD.

1. Observation of the Meissner Effect at T>Tc

The enhanced critical temperature Tc* was tested by making use of the two essential properties associated with superconductivity, namely: (1) diamagnetism, also known as the Meissner effect, which produces the so-called floating magnet phenomenon. The fact that no diamagnetism has ever been observed in YBCO except when it is in the superconducting state, strongly suggests that when IR irradiation on the YBCO disc surface produces diamagnetism, it has induced superconductivity; (2) zero resistance. To observe and calibrate the deviation or repulsion of the magnet from its lowest position above the non-superconducting sample disk due to Meissner effect, a 5 mm cube magnet was suspended as a pendulum about 3 mm above the YBCO tablet. Both tablet and pendulum were situated within the cryostat setup, which simply consisted of one inner glass jar nested in a second glass jar that held the liquid nitrogen. A type T thermocouple was used to measure the temperature in the inner jar with the YBCO lying flat on the bottom of the glass jar. All temperature data was recorded in degrees Celsius by a Fluke multi-meter. See Figure 5.1 which shows a schematic of the setup for experiment 1.

For the incoherent light source of wavelength 1300 nm [5], a hand-made set of Marubeni-Type C Infrared LEDs were used. All visual data of the Meissner effect throughout the temperature range, from liquid nitrogen temperature to room temperature, obtained from vaporization of the liquid nitrogen then followed by natural warming, was recorded with a Sony Handicam video camera model DCRTRV 19 (small camera) and the complete cryostat setup was projected on to a TV screen. A second camera Panasonic Digital3 CCD DVC Pro HD P2 (large camera) was used to record the data on the TV screen simultaneously with the data string (temperature at tablet; temperature in inner jar) shown on the Fluke meter RMS 87V and Fluke meter RMS 189, respectively. The schematic of the setup is shown in Figure 5.1.

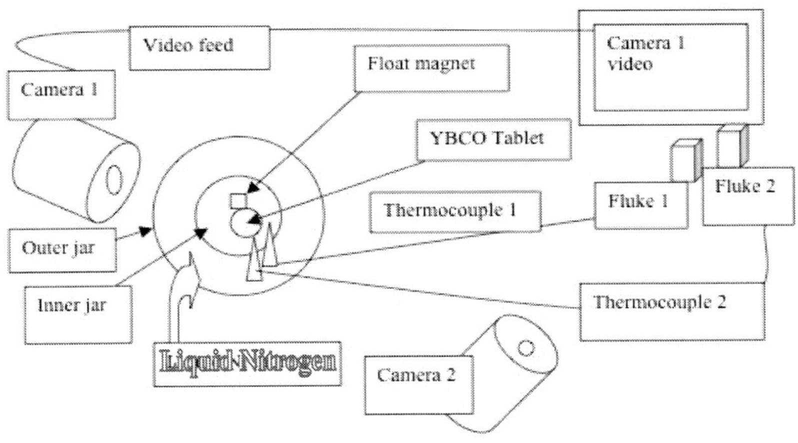

Figure 5.1. KansasIII Experiment 1: Meissner Effect setup schematic [5].

2. R-T Measurement in the Superconducting State at T>Tc

The enhanced critical temperature Tc* was further tested by using a four point probe to measure the resistance of the YBCO as a function of temperature. The verification of zero resistance in a large bulk sample is not easy, because YBCO is a relatively good conductor in its normal state and its total resistance is quite small. At room temperature our bulk sample has a resistivity of 20 mOhm. Thus, many factors can obscure our dc resistance measurement. For example, the accuracy of the voltmeter, the presence of Eddie currents, the potential presence of vortex dynamics, etc. Nevertheless, a marked reduction in the measured resistance of the sample from its normal phase before the onset of diamagnetism to the re-disappearance of diamagnetism still serves as a convincing indication that the sample is superconducting and its dc resistance is actually zero when diamagnetism was present. The experimental setup consisted again of the two nested jars with the YBCO resting on the bottom of the inner jar with the four-point probe soldered onto its surface. As shown in the schematic of Figure 5.4 and the image in Figure 5.5, the two outer wires are designed for applying current, the inner wires are for measuring the voltage drop. In addition, a thermocouple was located close to the center of the tablet between the 4-point probe contacts, for more precise measurement of the tablet temperature, and a second thermocouple measures the temperature of the inner jar. The current applied was 0.5 Amperes by making use of a dc-power supply operating in the current limited mode built by Ellison Research Labs [24]. The digital micro-voltmeter that was used to measure the voltage drop was a Agilent model 34420A with a resolution of 10 micro-volts. The incoherent light source of wavelength 1300nm [5], is a hand-made set of Marubeni-type C Infrared LEDs. The general setup is shown in Figure 5.2 and 5.4

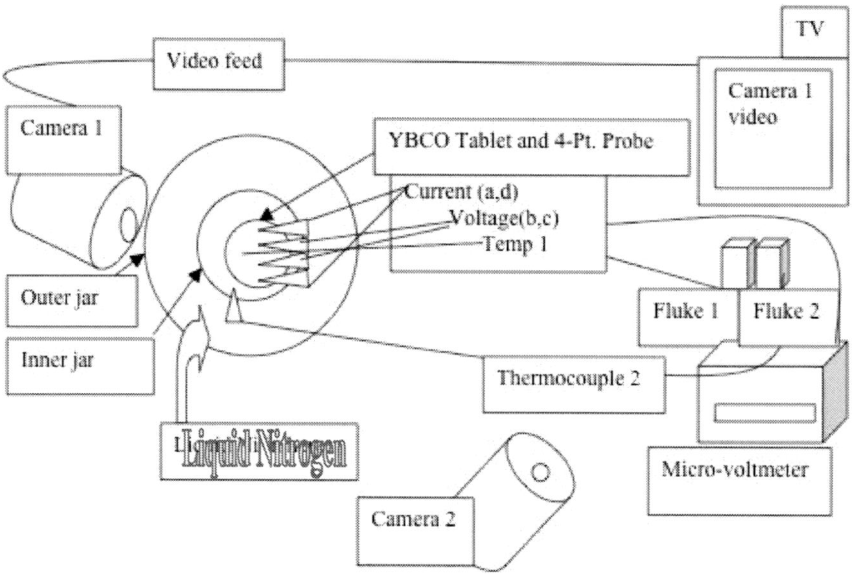

Figure 5.2. Experiment 2 setup schematic [5].

Figure 5.3. Experiment 2 Open system off-center-nested cryogenic-chamber showing 4-point probe and thermocouple on bulk YBCO surface [5]

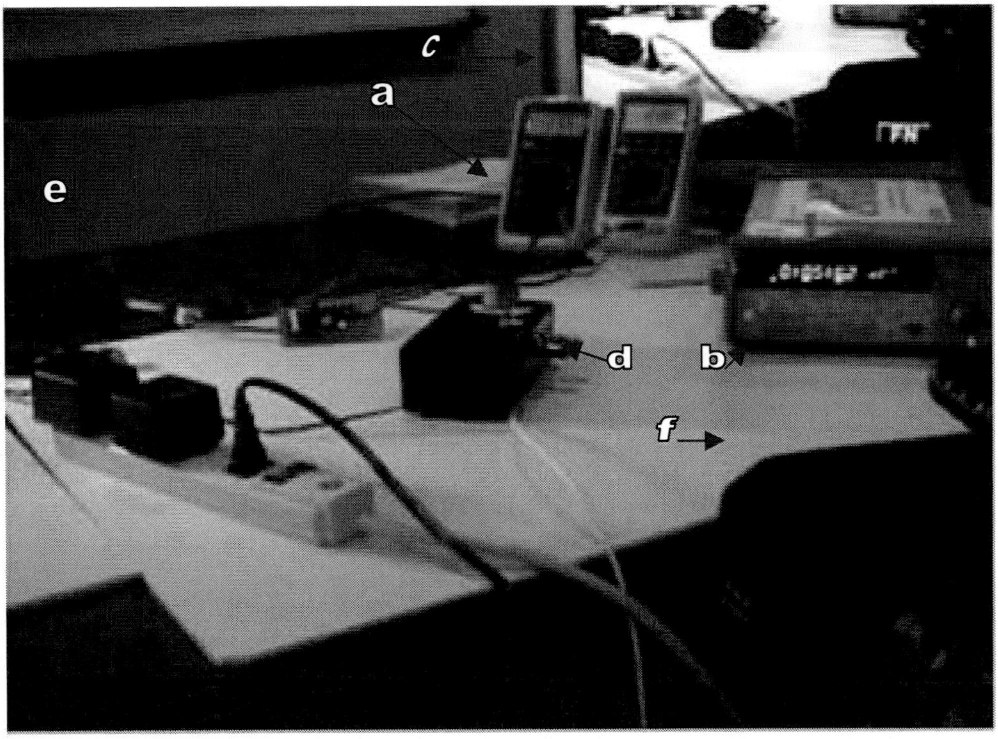

Figure 5.4. Test measurement setup showing: a. Fluke meters; b. Agilent microvoltmeter; c. TV projecting small camera view; d. dc-current limited power supply, e. IR light source with controller, f. large camera films all the data and TV image [5]

3. Tc Enhancement by Optical Pumping

The experimental results of Tc enhancement in the bulk YBCO disk by IR light irradiation can be summarized as follows:

Experiment (1). Using the pendulum-magnet the Meissner effect was observed in the range from 172.5°K to 190°K upon irradiating the YBCO tablet with the IR light source starting at 155.5°K. The initial onset of the magnet moving off the tablet is observed at approximately 157°K, and full deflection occurs at a temperature reading of Tc* = 172.5°K. The IR light source was kept on, with the magnet in its deflected position and remaining at such until the temperature of 190°K as indicated by the Fluke thermometer located on the inner jar wall was reached. At which point, the pendulum started to move back to its equilibrium position indicating the loss of the optimum exciton population necessary for superconducting conditions to be maintained, as shown in Figure 5.5 and 5.6. This large Tc* value beyond the EEM prediction could be due to the temperature difference inside the bulk disk as compared to that on the jar wall. In natural warming we do expect the temperature inside the bulk disk to be lower, as thermal equilibrium was not reached. However, we believe this difference is within 20°K, thus, in agreement with the EEM predicted value.

a b

Figure 5.5. Experiment 1 Onset (a) & final (b) position of magnet upon IR application [5].

Experiment (2). We believe greater detailed information can be gained about the onset of the enhanced superconducting state in YBCO by measuring resistance as a function of temperature. As seen clearly in Figure 5.7 the re-introduction of the superconducting state measured as zero Resistance occurs in the temperature range of about 176°K to 185°K upon application of IR light. This result was achieved first by cooling the tablet to near liquid nitrogen temperature. The tablet was then allowed to warm up to about 145°K and out of superconducting phase, at which point IR light was applied. This was followed by initial voltage oscillations induced by the IR as read by the Agilent microvoltmeter and cease once the optimum exciton population was reached as predicted by Wong [16] [5]. Should we have used a dc voltage source plus lock-in-amplifier, much of the large voltage fluctuations that were due to exciton number changes from IR photon absorption and emission will be

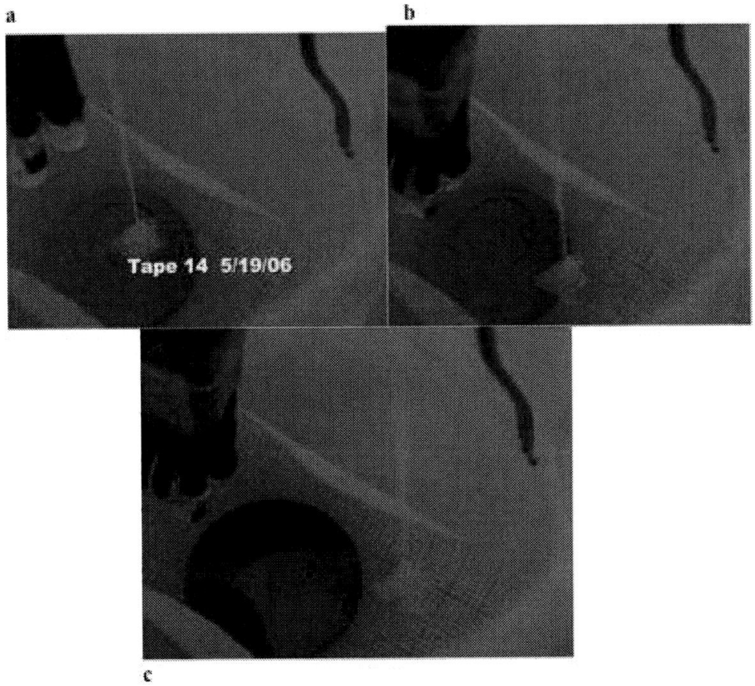

Figure 5.6. Experiment 1 Onset (a), half way (b) and final (c) position of magnet upon IR application [5]

screened-out; while the voltage from the normal phase resistance, as shown in Figure 5.7, from 189°K to 194°K, should remain. Thus, the dc source without a lock-in-amplifier does have an advantage of explicitly showing the IR effect on the exciton number. Obviously, the dc voltage source also creates inaccuracy in the resistance measurement. These fluctuations subsided at approximately 176°K and zero resistance. Ultimately the Tc* value was reached close to Tc* = 185°K (see figure 5.7) as voltage fluctuations reappeared. At a temperature of 189°K the IR light source was removed, but the superconducting state of the YBCO was not lost immediately until 200°K. We believe the reappearance of zero resistivity after IR light was removed at 189°K could be due to excited exciton decay in the sample, leading to actually a lowering of the inside temperature to below Tc*, thus, reproduces superconductivity. While it only requires the existence of a superconducting closed loop anywhere inside the sample to have zero resistance, it requires a substantial volume of superconducting material to be able to repel a magnet. Thus such a re-appearance in superconductivity was not seen in experiment 1. Eventually the excited exciton population density will become too low and superconductivity then disappeared. See Figure 5.7.

As clearly seen in Figure 5.7, the superconducting state at 176°K to 185°K is in agreement with the Meissner Effect observed in the pendulum-magnet experiment (1) from 173°K to 190°K. Regardless of the temperature measurement being within a few degrees K, the temperature enhancement above Tc of more than 80°K by optical stimulation of YBCO using IR incoherent light has been achieved and demonstrated to occur by employing two independent methods. The results in experiment (2) clearly are in agreement with the magnetic repulsion observed in experiment (1). Hence, we can conclude that, by making use of only very basic experimental components we have shown beyond a doubt that IR irradiation enhances Tc from 93°K to nearly 185°K $\pm 20°K$. This higher Tc* value as measured on the sample surface was primarily due to the thermal in-equilibrium inside the sample, caused by the natural warming process as compared to EEM predicted value of 175°K for such an optimum bulk sample.

6. Conclusion

The superconductivity state of a material is verified by simultaneous measurement of zero resistivity and diamagnetism, otherwise known as the Meissner effect. Normal experimental measurement of diamagnetism is employing an inductance coil over the superconductor and directly measuring the magnetic susceptibility. The enhanced Tc* superconducting state requires an IR irradiation directly onto the superconductor, thus it is not feasible for us to employ the above method to measure the magnetic susceptibility and still be able to irradiate the sample with IR light simultaneously. Because of the above difficulty we were forced to employ the open system method as described in the section 5. This method though far less accurate and desirable in determining the Tc* value, none-the-less, gives us an unmistakable marking of the presence of the Meissner effect. Tc* value can be accurately determined by

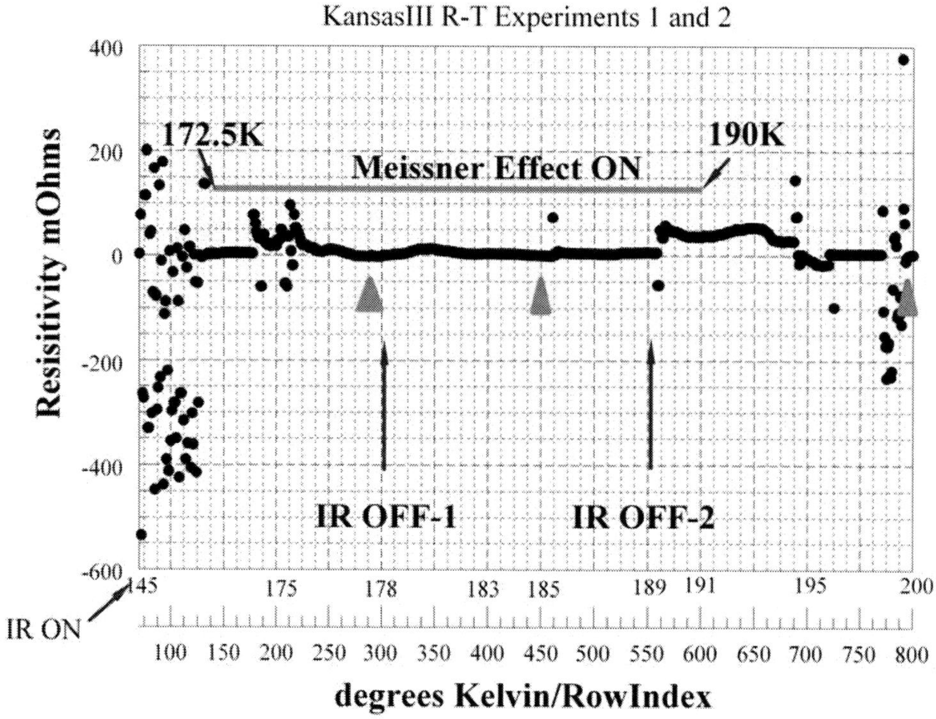

Figure 5.7. KansasIII R-T data set for experiments 1 and 2. After exiting Tc, IR irradiation of bulk sample starts at 155°K for Exp1 and 145°K (row 70) for Exp2. It shows: Exp1-Maissner Effect range: 172°K-190°K, Exp2-Tc* range: 176°K-185°K (rows 250-450). Following IR-OFF-1=178°K(row 300) and IR-OFF-2=189°K(row 550), we observe an additional Tc* at near 200°K (row 797) [5]

the closed cryogenic system chamber having an IR transparent window as described in section 4. As resistivity is a surface property, the resistivity is best measured on thin film samples, as IR light penetration has a finite penetration depth and thus the sample superconducting state is sample thickness dependent. Furthermore, as IR irradiation populates the density of the excited exciton (E_m), the natural decay of the excited states on the surface also affects the excited exciton population. Thus in the IR irradiation process, the sample needs to achieve its equilibrium excited exciton density (critical density) before Tc* can be accurately obtained. For thin film samples this equilibrium state can be reached in a relatively short time. The same cannot be said for a bulk sample. Last but not least, is the condition of the sample itself. For epitaxial thin films, due to lattice mismatch to the film's substrate it is common for such films to be more oxygen deficient, which naturally leads to reduced Tc value. The YBCO film we used in the experiment was oxygen deficient and has a Tc of only $69^{\circ}K$, far lower than the known optimum Tc of $93^{\circ}K$. For this reason, as shown by our EEM theory, the exciton binding energy was changed through the effective charge of the hole states in the exciton. We have carefully calculated this correction in section 3 (equations 26-29), and predicted the Tc* value for such a sample. This value is in exact agreement with the measured value. Although, as we mentioned earlier, the Tc* obtained through an open system experiment is not accurate. None-the-less, we obtained, a Tc* value on the YBCO bulk sample-with a nearly optimum Tc of $93^{\circ}K$ obtained through careful closed system cryogenic resistivity measurement [23], which is very close to the theoretical value of $175^{\circ}K$. Clearly, in the open system method and with the warming process we employed, there could exist a temperature gradient from the bulk sample core to its surface. The maximum error difference due to this temperature gradient in the bulk sample was determined to be less than $20^{\circ}K$ as caused by the natural warming process. Thus, the Tc* of 185°K and 190°K obtained for experiment 2 and experiment 1 respectively, is a clear indication of the effect of the IR light on enhancing Tc, irrespective of any inaccuracies of the method employed.

To conclude this paper, we would like to suggest a couple of new experiments: The R-T data from the thin film experiment clearly showed IR irradiation in the normal phase of the cuprate increases the charge carrier density. Since the charge carrier is composed of negative electrons and positive holes with unequal effective masses, we would expect this carrier density increase would be reflected in the thermal electric power measurement. A significant change of the thermal electric power might be useful for the design of a solid state cooler above Tc* temperatures. The second experiment we would like to propose is, encasing a cuprate conductor inside an optical fiber. This optical fiber must be totally transparent to IR light. Employing such an optical fiber cable, we might be able to enhance the Tc of the cable by passing the IR light of the correct frequency through, and thus reducing the needed cooling to achieve superconductivity in the cuprate wire. Neither of these suggested experiments is easy, but they do have useful application potential.

Acknowledgements

We would like to thank Dr. G. A. M. Dreschhoff for her valuable and tireless assistance in the collection of KansasIII and GenevaIII data, analysis and discussion and overall support regarding the EEM research. We would like to thank Dr. Dirk Van der Marel and Mr. Eric Van Heumen of the Condensed Matter Physics Department of Geneva University-Switzerland for providing their laboratory facilities and their valuable and tireless assistance in setup, equipment modification and the collection of the Geneva III data for EEM. And finally, a special thanks to Dr. Jan Zaanen of Leiden University- The Netherlands for listening to us and giving us a chance to go to Geneva and make it happen, when no one else would.

References

[1] K.W. Wong and W.Y. Ching, *Physica* **C416** (2004) 47
[2] D.R. Huffman, *Physics Today*, No.**11**, (1991) 22
[3] J.G. Bednorz, K.A. Muller, *Z.. Phys.* **B64** (1986) 188; C.W. Chu, et al, *Phys. Rev. Lett.* **58** (1987) 405
[4] Robert J. Cava, *Nature*, Vol.444, No.23 (2006) 427
[5] S. Curatolo and K.W. Wong, Method for operating a superconductor at a temperature higher than the superconductive temperature Tc, CZT Inc. *PCT/US06/62600*
[6] J.G. Bednorz, K.A. Muller, *Z. Phys.* **B64** (1986) 188
[7] R.F. Curl and R.E. Smalley, *Scientific American*, No.**4**, (1991) 54
[8] E. Bustarret et al, *Nature*, Vol.444, No.23, (2006) 465
[9] J. Bardeen, L.N. Cooper, J.R. Schrieffer, *Phys.Rev.***108** (1957) 1175
[10] J. Kolacek and P. Vasek, *Physica* **C336**, (2000) 199
[11] Y. Xin, K.W. Wong, C.X. Fan, Z.Z. Sheng and F.T. Chan, *Phys. Rev.* **B45** (1993) 557
[12] W.Y. Ching, Y. Zu, G.-L. Zhao, K.W. Wong, F. Zandiehnadem, *Phys. Rev. Lett.* **59** (1987) 1333
[13] H.L. Ji and K.W. Wong, *Phys. Lett.* **A256** (1999) 66
[14] J.R. Kirtley et al. *Nature* Vol. 373 (1995) 225.
[15] W.Y. Ching, Yongnian Xu and K.W. Wong, *Phys. Rev.* **B40** (1989) 7584
[16] K.W. Wong, On the question of Tc enhancement of high temperature superconductors, CZT Report [unpublished], October 2006
[17] R.L. Davidchack, K.W. Wong and P.C.W. Fung, *Phys. Lett.* **A223** (1996) 289
[18] Dr. Dirk Van Der Marel-Director, D.P.M.C., Departement de Physique de la Matiere Condensee. University of Geneva, 24 Quai Ernest-Ansermet. CH-1211-Geneva 4. Switzerland-
[19] THEVA Dünnschichttechnik GmbH,Rote-Kreuz-Str. 885737 Ismaning, Germany
[20] JANIS Research Company Inc. 2 Jewel Drive, P.O. Box 696 Wilmington, MA 01887-0696 USA
[21] Stanford Research Systems, Inc.1290-D Reamwood Ave., Sunnyvale, CA 94089 USA
[22] At the D.P.M.C., Departement de Physique de la Matiere Condensee. University of Geneva. 24 Quai Ernest-Ansermet. CH-1211-Geneva 4. Switzerland-photo showing Mr. Eric Van Heumen assisting.
[23] Colorado Superconductor Inc. –subsidiary of DRB Inc. Colorado Springs CO USA
[24] Ellison Research Labs Inc. *Wagon wheel drive.* Lawrence KS USA
[25] Robert F. Service, *Science* **310** (2005) 1271

In: Progress in Superconductivity Research
Editor: Oliver A. Chang, pp. 79-102

ISBN: 1-60021-694-3

Chapter 3

SPONTANEOUSLY BROKEN SYMMETRY AND INHOMOGENEOUS SUPERCONDUCTING STATES IN FERROMAGNET/SUPERCONDUCTOR NANOSTRUCTURES

Mansur G. Khusainov [1-3,*], ***Marat M. Khusainov*** [4] ***and Yurii N. Proshin*** [1,3,**]

[1)] Kazan State University, Kazan 420008, Russia
[2)] Kazan State Tupolev Technical University, Chistopol' 422981, Russia
[3)] Max-Planck-Institute for the Physics of Complex Systems, Dresden 01871, Germany
[4)] Zavoisky Physical-Technical Institute of RAS, Kazan 420029, Russia

Abstract

We derive a new boundary-value problem for the ferromagnetic metal/superconductor (FM/S) nanostructures, where superconductivity is a superposition of the BCS pairing with zero total momentum in the S layers and the Fulde-Ferrell-Larkin-Ovchinnikov (FFLO) pairing with nonzero 3D coherent momentum **k** in the FM layers. It is shown that processes of mutual transformation between the BCS and FFLO pairs at the S/FM boundary go via the unrealized *virtual states* (FFLO and BCS, correspondingly). These processes occur as the *Umklapp processes* during which excess coherent pair momentum **k** becomes confined in the FM layer. This fact is reflected in the significantly new boundary conditions for the Eilenberger equations. On this base we originally investigate the interplay between the BCS and 3D FFLO states in the pure FM/S bilayers. In the FM/S/FM trilayers two novel π *phase superconducting states with electron-electron repulsion* in the FM layers are predicted. The 3D modulated FFLO states are possible in such trilayers only in presence of external magnetic field at suitable parameters of the FM and S layers. In the FM/S superlattices there are also two π phase magnetic states (0π and $\pi\pi$) with compensation of the exchange field paramagnetic effect. This fact allows us to explain a surprisingly high $T_c \sim 5$K in the short

[*] E-mail address:mgkh.kgtu@mail.ru
[**] E-mail address: Yurii.Proshin@ksu.ru

period Gd/La superlattice and to predict the sign and value of the electron-electron interaction in the Gd metal.

Keywords: superconductivity, ferromagnetism, proximity effect, mutual accommodation, boundary value problem, Umklapp processes

PACS: 74.80.Dm; 74.50.+r; 74.62.-c

1. Introduction

In recent years there is a pronounced interest in unconventional superconductivity with superconducting electronic correlations different from the usual singlet BCS pairing with zero total momentum. One such example is the Fulde-Ferrell-Larkin-Ovchinnikov (FFLO) type pairing with a nonzero momentum of pairs [1,2], which may be realized not only in a ferromagnetic superconductor but also in the layered ferromagnetic metal/superconductor (FM/S) structures (see reviews [3,4] and references therein). The competition of superconducting and magnetic states in the layered FM/S structures leads to the pronounced non-monotonous dependence of the critical temperature T_c on the thickness of the FM layers d_f [3,4]. The existing theories of the proximity effect [3,4] connect the nature of non-monotonous behavior of $T_c(d_f)$ with vibrations of the Cooper pairs flux at the FM/S boundary, which arise due to one-dimensional (1D) oscillations of the pair amplitude across the FM layers. However, these theories predict multiple oscillations of $T_c(d_f)$ or even periodically reentrant superconductivity. Whereas, only one local maximum (or minimum) in the $T_c(d_f)$ dependence in the majority of experiments with real 3D multilayers Fe/V or Gd/Nb is realized. The reason is that the former 1D theories [3] neglect by spatial changes of pair amplitude along the FM/S boundary. Moreover, the recently observed absence of the 3D superconductivity suppression in the short period Gd/La superlattice [5] is a real "physical challenge" to the existing proximity effect theory. Usually the superconductivity in analogous FM/S layered structures takes place if only the FM layer thickness d_f is much less than the S layer critical thickness d_{sc}, which must be of order of the coherence length ξ_s due to the strong paramagnetic effect of exchange field I in the ferromagnet [3,4]. The typical examples of above-mentioned behavior are the Gd/Nb and Fe/V superlattices. Whereas in the case of the Gd/La superlattice [5] $d_s \ll \xi_s$ and moreover $d_s \ll d_f$ nevertheless superconductivity not only arises but it arises at the critical temperature T_c corresponding to the "bulk" lanthanum value that is surprisingly high for a heterogeneous system. Below we will show that full boundary-value problem have solutions which explain the Gd/La surprising behavior

Up to now, beginning with the pioneer works of Radović et al., Buzdin et al. and in the further papers (see references in [3,4]), it was considered that pair amplitude $F(\mathbf{r})$ can change only across the FM and S layers. This corresponds to the 1D state with $\mathbf{q}_f = 0$ where $\mathbf{q}_f$ is the 2D (in the FM/S boundary plane) wave vector of the FFLO pairs. The possible realization of only the 1D state with $\mathbf{q}_f = 0$ was reasoned by Demler, Arnold and Beasley [6] by conservation of the tangential component of the pairs momentum at the FM/S boundary. In the dirty limit such proposal leads to the 1D boundary conditions for pair amplitude at the FM/S interface, which we derived earlier [3]. Unfortunately, these 1D conditions are not adequate to the 3D FFLO pairing nature in the FM layers. Below we will show that new 3D states have the higher T_c and therefore they realize rather than the old 1D states. Thus the superconductivity in the FM/S systems proves to be a superposition of the BCS pairing in the S layers and the 1D-3D FFLO pairing scenario in the FM layers. Investigation of the mutual accommodation of these competing types of pairing is the main goal of the present paper.

2. Integral Boundary-Value Problem for FM/S Bilayer

Let us consider the planar contact between the FM layer ($-\infty < z < 0$) and the S layer ($0 < z < \infty$). The critical temperature T_c of the inhomogeneous superconductor is defined by the Gor'kov self-consistency equation [7] for the order parameter $\Delta(\mathbf{r})$

$$\Delta(\mathbf{r}) = 2\lambda(z)\pi T \operatorname{Re} \sum_{\omega>0}{}' F(\mathbf{r},\omega). \tag{1}$$

Here $F(\mathbf{r},\omega) = F(\mathbf{r},\mathbf{r},\omega)$ is the one-point anomalous Gor'kov function; $\lambda(z>0) = \lambda_s$ and $\lambda(z<0) = \lambda_f$ are the dimensionless parameters of electron-electron interaction, and prime at the sum sign means a cutoff on the Debay frequency ω_D; $\omega = \pi T(2n + 1)$ is the Matsubara frequency, T is the temperature, and hereinafter in paper $\hbar=k_B=\mu_B=1$. The anomalous Gor'kov function $F(\mathbf{r},\omega)$ is connected with the correlation function $H(\mathbf{r},\mathbf{r}',\omega)$ by the integral equation

$$F(\mathbf{r},\omega) = \frac{1}{\pi N(z)} \int H(\mathbf{r},\mathbf{r}',\omega)\Delta(\mathbf{r}')d^3r', \tag{2}$$

where the kernel of Eq. (2) is defined by relation

$$H(\mathbf{r},\mathbf{r}',\omega) = \left\langle G^0{}_{\uparrow}(\mathbf{r},\mathbf{r}',\omega)G^0{}_{\downarrow}(\mathbf{r},\mathbf{r}',-\omega)\right\rangle_{imp}. \tag{3}$$

Here $G^0{}_{\uparrow,\downarrow}(\mathbf{r},\mathbf{r}',\omega)$ is the electronic Green function in the normal state with spin indices $\uparrow,\downarrow$, and the angle brackets mean the averaging over all impurities configurations.

By diagram technique [7] we can show that in the presence of the exchange field and nonmagnetic impurities scattering the kernel $H(\mathbf{r},\mathbf{r}',\omega)$ of Eq. (2) is the solution of another integral equation

$$H(\mathbf{r},\mathbf{r}',\omega) = K(\mathbf{r},\mathbf{r}',\omega) + \int \frac{K(\mathbf{r},\mathbf{r}_1,\omega)H(\mathbf{r}_1,\mathbf{r}',\omega)d^3r_1}{2\pi N(z_1)\tau(z_1)}. \tag{4}$$

Here $N(z)$ and $\tau^{-1}(z)$ are the density of states on the Fermi surface and the rate of the electron-impurity scattering, respectively, which step-wise change their values when crossing the FM/S boundary z=0. The kernel $K(\mathbf{r},\mathbf{r}',\omega)$ of Eq. (4) is defined by the following expression

$$K(\mathbf{r},\mathbf{r}',\omega)=G_\uparrow(\mathbf{r},\mathbf{r}',\omega)G_\downarrow(\mathbf{r},\mathbf{r}',-\omega), \tag{5}$$

where $G_{\uparrow,\downarrow}(\mathbf{r},\mathbf{r}',\omega)$ is the conduction electrons Green function in the normal phase averaged over-all the impurity configurations. The diagrammatic representation of the integral equation (4) is given by Fig. 1.

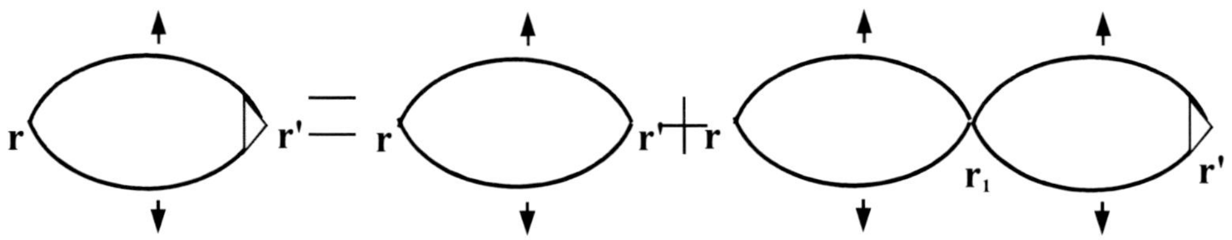

Figure 1. Diagrammatic representation of the Eq. (4). Shaded loop presents the vertex, which describes the correlation function $H(\mathbf{r},\mathbf{r}',\omega)$, solid lines correspond to the Green functions $G_{\uparrow,\downarrow}(\mathbf{r},\mathbf{r}',\omega)$. The each simple loop corresponds to the kernel $K(\mathbf{r},\mathbf{r}',\omega)$. Over common for two loops radius-vector $\mathbf{r}_1$ it is suggested the integration normalized on the $2\pi N(z_1)\tau(z_1)$ in the case of the flat FM/S boundary at $z_1=0$. Vertical arrows: ↑, ↓ denote the spin orientations.

Eq. (4) unlike to analogous integral equation of our previous paper [8] is three-dimensional and together with the expressions (1), (2) allows us to take into account the spatial variations of the order parameter and pair amplitude not only across the FM and S layers (along z axis), but in plane x–y of the FM/S boundary also. This fact gives hope to receive thereafter more wide class of the T_c (d_f) behaviors than previously (see Refs. [3,4]).

By substitution of Eq. (4) in definition (2) we easily derive the integral boundary-value problem for the Gor'kov function $F(\mathbf{r},\omega)$ in the FM/S contact near T_c

$$F(\mathbf{r},\omega)=\frac{1}{\pi N(z)}\int K(\mathbf{r},\mathbf{r}',\omega)\left[\Delta(\mathbf{r}')+\frac{1}{2\tau(z')}F(\mathbf{r}',\omega)\right]d^3r'. \tag{6}$$

Integral equation (6) together with self-consistency Eq. (1) contain full information about the electron-electron interactions, parameters of electronic structure and kinetic characteristics of contacting metals, as well as about step-wise change of their values when crossing the sharp FM/S separation boundary. However, for solution of the Eq. (6) in general case it is necessary to know the kernel $K(\mathbf{r},\mathbf{r}',\omega)$, which can be found from Eq. (5) as solution of the corresponding differential boundary value problem for the normal Green functions $G_{\uparrow,\downarrow}(\mathbf{r},\mathbf{r}',\omega)$ in the FM/S contact. Thus in general case the boundary value problem for $F(\mathbf{r},\omega)$ is very difficult. While at the quasiclassical approach, in which the spatial oscillations are neglected on atomic scale, it is more convenient to reduce the problem of integral equation (6) solving for the $F(\mathbf{r},\omega)$ in the FM/S contact to the solution of the equivalent differential boundary value problem (see Sec. 4).

Below in the Sec.3 we will find the simplest solutions of the integral boundary value problem (6) in the case of uniform bulk ferromagnetic superconductor in order to investigate the competition between the usual BCS and unconventional FFLO states in the framework of the Eilenberger equations [9] for the anomalous Green function.

3. Competition between BCS and FFLO States in Ferromagnetic Superconductor

In the case of the infinite homogeneous ferromagnetic superconductor due to translation invariance of system the kernel $K(\mathbf{r},\mathbf{r}',\omega) = K(\mathbf{r}-\mathbf{r}',\omega)$ depends on the modulus of difference $R = |\mathbf{r} - \mathbf{r}'|$ only. In the quasiclassical limit [7] this dependence for pure ferromagnet (with $\tau_f^{-1} = 0$) has the pronounced 3D oscillating character

$$K(\mathbf{r},\mathbf{r}',\omega) = \left(\frac{m}{2\pi R}\right)^2 \exp\left(-\frac{2\omega + 2iI}{v_f}R\right), \tag{7}$$

where m is the electron mass, I is the exchange field and v_f is the Fermi velocity. Let us search for the solution of Eq. (6) in the simplest case with one Fourier harmonics [10]

$$\Delta(\mathbf{r}) = \Delta_\mathbf{k} \exp(i\mathbf{kr}), \tag{8}$$

where $\mathbf{k}$ is the coherent pairs momentum. Then by substitution of the kernel (7) and the order parameter (8) into Eq. (6) it is easy to show that the last one is reduced to equation

$$F(\mathbf{r},\omega) = \oint \frac{d\Omega_\mathbf{p}}{4\pi} \Phi(\mathbf{p},\mathbf{r},\omega) = \langle \Phi(\mathbf{p},\mathbf{r},\omega) \rangle_{\Omega\mathbf{p}}, \tag{9}$$

where $\Phi(\mathbf{p},\mathbf{r},\omega)$ is the Eilenberger function [9], $\mathbf{p}$ is the Fermi momentum and the circle integral denotes integration on full space angle of the Fermi sphere. The Eilenberger function, entering in the Eq. (9), has the form

$$\Phi(\mathbf{p},\mathbf{r},\omega) = \frac{2\Delta_\mathbf{k} \exp(i\mathbf{kr})}{\left[2\omega + i(2I + \mathbf{v}_f \mathbf{k})\right]}, \tag{10}$$

where $\mathbf{v}_f$ is the Fermi velocity vector. As it follows from Eq. (10) the function $\Phi(\mathbf{p},\mathbf{r},\omega)$ satisfies to the linear differential equation

$$\left[2(\omega + iI) + \mathbf{v}_f \nabla_\mathbf{r}\right] \Phi(\mathbf{p},\mathbf{r},\omega) = 2\Delta(\mathbf{r}), \tag{11}$$

that coincides with corresponding Eilenberger's equation linearized near T_c [9]. We can see that the pole of Eilenberger function (10) contains pair-breaking factor $i(I + \mathbf{v}_f\mathbf{k}/2)$, which might tend to zero at some values of $k \sim 2I/v_f$ and cosine of angle between $\mathbf{v}_f$ and $\mathbf{k}$. That is why the FFLO state originates at defined values of the exchange field I and becomes even energetically favorable than the BCS state.

Thus the integral equation (6) with kernel (7) and the equivalent set of Eqs. (9)-(11) in case of the pure uniform ferromagnetic superconductor have two solutions of the different symmetry. First solution is the BCS type with $\mathbf{k}$ = 0, at which $\Delta(\mathbf{r})$ and $F(\mathbf{r},\omega)$ are not $\mathbf{r}$ dependent was found by Gor'kov and Rusinov [11]. Second solution is the 3D oscillating FFLO state, which in the simplest case look like $F(\mathbf{r},\omega) \propto \Delta(\mathbf{r}) \propto \exp(i\mathbf{kr})$ [1,2], where the coherent momentum of the FFLO pairs $k \approx 2I/v_f$ at $2I << \varepsilon_F$ (ε_F is the Fermi energy). Let us substitute the Eqs. (9), (10) into Eq. (1) and after summation over ω we easily obtain that the full solution for the reduced critical temperature $t = T_c/T_{c0}$ of the pure uniform ferromagnetic superconductor can be find from following equation

$$\ln t = \Psi\left(\frac{1}{2}\right) - \frac{1}{2}\mathrm{Re}\int_{-1}^{1} dx \Psi\left(\frac{1}{2} + \frac{i(2I + xkv_f)}{4\pi T_{c0} t}\right). \tag{12}$$

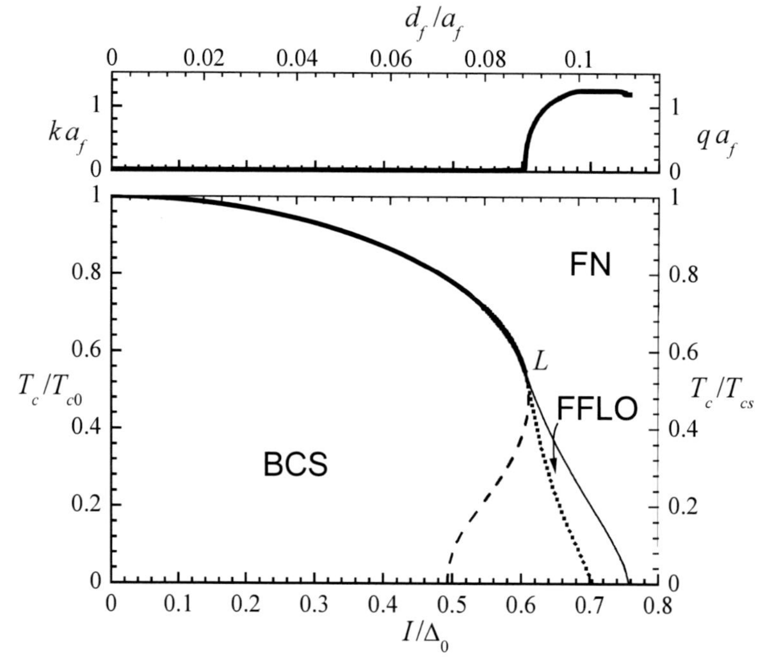

Figure 2. The phase diagrams (T_c – I) – lower panel and (k – I) – upper panel for the 3D uniform ferromagnetic superconductor. In this case only the left and lower axes are working. The thick solid line is the line of second order phase transitions between the BCS and normal states. The thin solid line below the Lifshitz point L is the line of second order phase transitions between the FFLO and normal states. The dashed line below L represents the unrealized virtual BCS state with $k = 0$. Finally the dotted line corresponds to the first order transitions [3,4] between the uniform BCS state with $k = 0$ and the inhomogeneous FFLO state with $k \neq 0$.

Here $\Psi(y)$ is the digamma-function, T_{c0} is the critical temperature of the FM defined by only the electron-electron interaction $\lambda_f > 0$ in the assumption of the absence of the exchange field I; x is the cosine of angle between the 3D pairs momentum $\mathbf{k}$ and the Fermi velocity $\mathbf{v}_f$. The concrete magnitude k is defined from the T_c maximum condition. The corresponding phase diagram: the reduced critical temperature t versus the normalized exchange field I/Δ_0 is well known [3,4]; it is depicted on the lower panel of Fig. 2. Here Δ_0 is the superconducting gap at $T = 0$. We can see that the BCS state with the uniform order parameter Δ and lower free energy is realized with increasing I up to the Lifshitz point L [12] with $t_L = 0.56$ and $I_L/\Delta_0 = 0.61$ on the second order phase transitions line. Whereas the FFLO state with the oscillatory Δ has a higher free energy and it exists as the unrealized *virtual state* up to the L point. After the Lifshitz point L the FFLO state already has a lower free energy and it takes place as the *ground state* in comparison with the unrealized *virtual* BCS state (dotted line).

The upper axes (d_f/a_f) for both panels and right axes (T_c/T_{cs} and qa_f) will be explained below in Sec. 6.

Note that our approach allows us to supplement this well-known picture by the dependence of the optimized reduced pair's momentum ka_f versus the same parameter I/Δ_0 (the upper panel of Fig. 2). Here $a_f = v_f/2I$ is the spin stiffness length in ferromagnetic metal. We can see that the value $k \neq 0$ at the first time arises in the Lifshitz point L when the FFLO state originates. At zero temperature the FFLO superconducting state is suppressed in the point $k_0 = 1.2/a_f$ by the paramagnetic effect of the critical exchange field ($I_c = 0.755\ \Delta_0$).

This Section has mainly a methodical sense. Indeed, in the usual superconductor S the BCS state only takes place as the ground state while the FFLO state always is virtual one. At the same time the ferromagnetic superconductor has two possible ground states: the smooth BCS state and the 3D spatially oscillating FFLO state. Therefore the question about continuous matching between the pair amplitudes of different spatial symmetry like the BCS and FFLO on the FM/S boundary promises to be very interesting.

4. Differential Boundary-Value Problem for the FM/S Contact

Let us return to the inhomogeneous FM/S superconducting system. It is important to note that the exchange field I, which step-wise tends to zero at transition from the FM layer to the S layer, breaks the translational symmetry at the FM/S boundary not only along z axis, but in the $\boldsymbol{\rho} = (x,y)$ plane also. In particular, the kernel $K(\boldsymbol{\rho},\boldsymbol{\rho}',z,z',\omega) \neq K(\boldsymbol{\rho}-\boldsymbol{\rho}',z,z',\omega)$, if $\boldsymbol{\rho}$ and $\boldsymbol{\rho}'$ belong to different sides of the FM/S contact (i.e. if $z{\cdot}z' < 0$). For example, if $\boldsymbol{\rho}$ varies inside the FM layer ($z < 0$), and $\boldsymbol{\rho}'$ varies inside the S layer ($z' > 0$), this kernel will be product monotone function of ($\boldsymbol{\rho} - \boldsymbol{\rho}'$) by oscillating one in $\boldsymbol{\rho}$. The equality $K(\boldsymbol{\rho},\boldsymbol{\rho}',z,z',\omega) = K(\boldsymbol{\rho} - \boldsymbol{\rho}',z,z',\omega)$ takes place only if both $\boldsymbol{\rho}$ and $\boldsymbol{\rho}'$ belong to the same layer (FM or S). We point to *spontaneous breakdown of translational symmetry* at the FM/S boundary *x-y* plane in contradiction to all previous papers on this theme. Therefore below we use separate Fourier transform in $\boldsymbol{\rho}$ and $\boldsymbol{\rho}'$, contrary to all former 1D theories of the proximity effect [3,4].

Since our problem (6) is linear in $F(\mathbf{r},\omega)$ and $\Delta(\mathbf{r})$, which are expressed in terms each other, we can use the Fourier expansion for solution of the Eq. (6). In the simplest case it is enough to consider the solutions with one Fourier harmonics [9]. Therefore we will search for the solutions for the $\Delta(\boldsymbol{\rho},z)$ and $F(\boldsymbol{\rho},z,\omega)$ in regions FM ($-\infty < z < 0$) and S ($0 < z < \infty$), correspondingly, in the form

$$\begin{aligned} &\Delta_f(\boldsymbol{\rho},z) = \Delta_f(\mathbf{q}_f,z)e^{i\mathbf{q}_f\boldsymbol{\rho}}; \quad \Delta_s(\boldsymbol{\rho},z) = \Delta_s(\mathbf{q}_s,z)e^{i\mathbf{q}_s\boldsymbol{\rho}}; \\ &F_f(\boldsymbol{\rho},z,\omega) = F_f(\mathbf{q}_f,z,\omega)e^{i\mathbf{q}_f\boldsymbol{\rho}}; \quad F_s(\boldsymbol{\rho},z,\omega) = F_s(\mathbf{q}_s,z,\omega)e^{i\mathbf{q}_s\boldsymbol{\rho}}, \end{aligned} \tag{13}$$

where the 2D components of pairs momentum $\mathbf{q}_f$ and $\mathbf{q}_s$ describe the possible oscillations of order parameter in the *x-y* plane of the FM/S bilayer. Here and below the index *s*(*f*) denotes the belonging of respective parameters and functions to the S(FM) metal. The solutions (13) possess satisfactory generality. Specifically, they include the previously studied 1D case $q_f = q_s = 0$, which corresponds to conservation of the tangential momentum of pairs at transition through the FM/S boundary [3,4,6]. In addition, these solutions allow the most dramatic competition between the BCS pairing in the S layer ($q_s = 0$) and the 3D FFLO pairing in the FM layer ($q_f \neq 0$). In latter case we deal with different types of spatial symmetry of order parameter in the FM and S regions.

Using the representation (13), the Eq. (6) it is convenient to rewrite in terms of the 2D Fourier components of corresponding functions, i.e.

$$F(\mathbf{q},z,\omega)=\frac{1}{\pi N(z)}\int_{-\infty}^{\infty}K(\mathbf{q},\mathbf{q}',z,z',\omega)\left[\Delta(\mathbf{q}',z')+\frac{1}{2\tau(z')}F(\mathbf{q}',z',\omega)\right]dz', \quad (14)$$

where 2D pairs momenta in the contact plane **q** and **q'** are the step wise-like functions of z and z', respectively. In particular, $\mathbf{q}(z) = \mathbf{q}_f\theta(-z) + \mathbf{q}_s\theta(z)$, where $\theta(z)$ is the step function; the dependence $\mathbf{q}'(z')$ has analogous form. Due to absence of the translational invariance, the kernel of the Eq. (14) can be written down as the production of the 2D Fourier representations of Green functions $G_\uparrow$ and $G_\downarrow$. This Fourier transform is performed separately on $\boldsymbol{\rho}$ and $\boldsymbol{\rho}'$. So, in case of the flat boundary FM/S, which conserves 2D projections of Fermi momenta ($p_{Fs}\sin\theta_s = p_{Ff}\sin\theta_f$; $\theta_{s(f)}$ is the angle between Fermi momentum $\mathbf{p}_{Fs(f)}$ and z axis) the kernel $K(\mathbf{q},\mathbf{q}',z,z',\omega)$ of Eq. (14) is expressed via the sum over $\mathbf{p} = \mathbf{p}'$, i.e.

$$K(\mathbf{q},\mathbf{q}',z,z',\omega)=\sum_{\mathbf{p}=\mathbf{p}'}\widetilde{K}(\mathbf{p},\mathbf{q},\mathbf{p}',\mathbf{q}',z,z',\omega), \quad (15)$$

$$\widetilde{K}(\mathbf{p},\mathbf{q},\mathbf{p}',\mathbf{q}',z,z',\omega)=G_\uparrow(\mathbf{p},\mathbf{p}',z,z',\omega)G_\downarrow(\mathbf{q}-\mathbf{p},\mathbf{q}'-\mathbf{p}',z,z',-\omega). \quad (16)$$

For the contact between the impure metals FM and S we can generalize our approach [3,8] on 3D case, i.e. via the solution of the problem with potential barrier at the FM/S boundary for the kernel $\widetilde{K}(\mathbf{p},\mathbf{q},\mathbf{p}',\mathbf{q}',z,z',\omega)$, which defined in (16). Then it can be shown in quasiclassical limit that

$$\begin{aligned}\tilde{K}_{ss}(\mathbf{p}_s,\mathbf{q}_s,\mathbf{p}_s,\mathbf{q}_s,z>0,z'>0,\omega)&=\frac{1}{v_{sz}^2}\left\{\exp\left(-\frac{|z-z'|}{\xi_{sz}}\right)+(1-\sigma)\exp\left(-\frac{(z+z')}{\xi_{sz}}\right)\right\};\\ \tilde{K}_{sf}(\mathbf{p}_s,\mathbf{q}_s,\mathbf{p}_f,\mathbf{q}_f,z>0,z'<0,\omega)&=\frac{\sigma}{v_{sz}v_{fz}}\exp\left(-\frac{z}{\xi_{sz}}+\frac{z'}{\xi_{fz}}\right).\end{aligned} \quad (17)$$

Here $\mathbf{p}_s = \mathbf{p}_f = \mathbf{p}$ for the flat boundary; and σ is quantum-mechanical transparency of barrier. In case if $z < 0$ and $z' < 0$ in the upper of Eqs.(17) the indices s are necessary replaced by $f\,(s \to f)$, and the signs of z and z' are changed for the opposite ones. If z<0 and z'>0, in the lower of Eqs.(17) it is necessary to interchange them ($z \leftrightarrow z'$). In the Eqs.(17) $v_z(z)$ and $\xi_z(z)$ are the z components of the Fermi velocity and the correlation lengths which are defined by following relations

$$\xi_z(z) = \frac{v_z(z)}{2\tilde{\omega}(z)}; \quad 2\tilde{\omega}(z) = 2\omega + \frac{1}{\tau(z)} + i[2I(z) + \mathbf{v}_{\perp}(z)\mathbf{q}(z)], \tag{18}$$

where $\mathbf{v}_{\perp}(z)$ is the projection of Fermi velocity $\mathbf{v}(z)$ on the x-y plane. The parameters which enter in Eqs.(12),(13) step-wise change their values at crossing the FM/S boundary $z = 0$, i.e. $I(z > 0) = 0$, $I(z < 0) = I$; $\tau(z > 0) = \tau_s$, $\tau(z < 0) = \tau_f$; $v(z > 0) = v_s$, $v(z < 0) = v_f$ etc. For the delta-like potential barrier $U\delta(z)$ at the FM/S boundary the transparency σ looks as

$$\sigma = \frac{4v_{sz}v_{fz}}{U^2 + (v_{sz} + v_{fz})^2}. \tag{19}$$

It is easy to check that the kernels (17) satisfy the sum rule

$$\int \tilde{\omega}(\mathbf{r})K(\mathbf{r},\mathbf{r}',\omega)d\mathbf{r} = \pi N(\mathbf{r}'),$$

which may be derived from sufficiently general arguments, see for example Ref. [14]. This means that all reflected and transmitted waves through the FM/S interface are correctly taken into account. Then we can restore the corresponding quasiclassical boundary-value problem, which has solution (17). This boundary value problem contains the differential equation

$$\left[2\tilde{\omega}(z) - v_z(z)\xi_z(z)\frac{\partial^2}{\partial z^2}\right]v_z(z)\tilde{K}(\mathbf{p},\mathbf{q},\mathbf{p}',\mathbf{q}',z,z',\omega) = 2\delta(z - z') \tag{20}$$

and corresponding boundary conditions at the flat FM/S boundary $z = 0$

$$v_z(z)\xi_z(z)\frac{\partial\tilde{K}(z,z',\omega)}{\partial z}\bigg|_{z=\pm 0} = \frac{\sigma}{2(1-\sigma)}\left[v_{sz}\tilde{K}(+0,z',\omega) - v_{fz}\tilde{K}(-0,z',\omega)\right], \tag{21}$$

where some arguments are dropped for brevity. In Eqs.(17)-(21) we have only $\mathbf{q} = \mathbf{q}'$ if $z{\cdot}z'$> 0, and on the contrary, if $z{\cdot}z'$< 0, we can have $\mathbf{q} \neq \mathbf{q}'$. At derivation of Eqs.(17)-(21) we neglect the fast oscillating and damping terms which are proportional to $\exp[\pm i p_F(z+z')]$, that corresponds to the quasiclassical approach $p_F l, p_F\xi >> 1$. Due to condition $2I << \varepsilon_F$ we neglect also the difference between electronic parameters belonging to dissimilar spin-subbands.

For following calculations it is convenient to introduce the Eilenberger function $\Phi(\mathbf{p},\mathbf{q},z,\omega)$ [11], which is connected with the one-point Gor'kov function $F(\mathbf{q},z,\omega)$ by relation

$$F(\mathbf{q},z,\omega)=\oint\frac{d\Omega_{\mathbf{p}}}{4\pi}\Phi(\mathbf{p},\mathbf{q},z,\omega)=\langle\Phi(\mathbf{p},\mathbf{q},z,\omega)\rangle_{\Omega\mathbf{p}}, \tag{22}$$

where the circle integral denotes integration on full space angle of the Fermi sphere. Then we can rewrite Eqs. (13),(14) in the form

$$\Phi(\mathbf{p},\mathbf{q},z,\omega)=\int_{-\infty}^{\infty}v_z(z)\widetilde{K}(\mathbf{p},\mathbf{q},\mathbf{p}',\mathbf{q}',z,z',\omega)\left[\Delta(\mathbf{q}',z')+\frac{1}{2\tau(z')}F(\mathbf{q}',z',\omega)\right]dz'. \tag{23}$$

Applying the boundary value problem (20), (21) to Eq.(23), it is easy to get following quasiclassical differential equations

$$\left[2\widetilde{\omega}(z)-v_z(z)\xi_z(z)\frac{\partial^2}{\partial z^2}\right]\Phi(\mathbf{p},\mathbf{q},z,\omega)=2\Delta(\mathbf{q},z)+\frac{1}{\tau(z)}F(\mathbf{q},z,\omega) \tag{24}$$

with related boundary conditions at the flat FM/S interface

$$\xi_z(z)\left.\frac{\partial\Phi(\mathbf{p},\mathbf{q},z,\omega)}{\partial z}\right|_{z=\pm 0}=\frac{\sigma}{2(1-\sigma)}\left[\Phi(\mathbf{p},\mathbf{q}_s,+0,\omega)-\Phi(\mathbf{p},\mathbf{q}_f,-0,\omega)\right]. \tag{25}$$

In these expressions the 2D projections of Fermi momenta $\mathbf{p}_s = \mathbf{p}_f = \mathbf{p}$ are conserved.

It is easy to prove that Eqs. (24) are reduced to the Eilenberger equations linearized near T_c [11]. Full solution of boundary value problem (24), (25) for the Gor'kov function $F(\boldsymbol{\rho},z,\omega)$ is rather complicated due to strong angle dependence of $\Phi(\mathbf{p},\boldsymbol{\rho},z,\omega)$. Nevertheless a main characteristic property of this boundary value problem can be understood. In particular, in the case of ideal transparency ($\sigma\rightarrow 1$) we get from Eq. (25) the continuity condition for the function $\Phi(\mathbf{p},\mathbf{q},z,\omega)$, i.e.

$$\Phi_s(\mathbf{p},\mathbf{q}_s,+0,\omega)=\Phi_f(\mathbf{p},\mathbf{q}_f,-0,\omega) \tag{26}$$

and the equality condition for fluxes of this function

$$\xi_{sz}\left.\frac{\partial\Phi_s(\mathbf{p},\mathbf{q}_s,z,\omega)}{\partial z}\right|_{z=+0}=\xi_{fz}\left.\frac{\partial\Phi_f(\mathbf{p},\mathbf{q}_f,z,\omega)}{\partial z}\right|_{z=-0}. \tag{27}$$

From the Eq. (26) after angle averaging and inverse Fourier transform we obtain the following continuity conditions ($\sigma\rightarrow 1$) for the Gor'kov functions (see also Eqs. (13))

$$\begin{aligned}F_s(\boldsymbol{\rho},+0,\omega)&=F_f(\boldsymbol{\rho},-0,\omega)\exp(-i\mathbf{g}\boldsymbol{\rho});\\ F_f(\boldsymbol{\rho},-0,\omega)&=F_s(\boldsymbol{\rho},+0,\omega)\exp(i\mathbf{g}\boldsymbol{\rho}),\end{aligned} \tag{28}$$

where $\mathbf{g} = \mathbf{q}_f - \mathbf{q}_s$ is the possible mismatch between 2D pairs momentum in the FM and S layers. In the case $q_s = q_f = g = 0$ the Eqs. (28) correspond to continuous matching on the S/FM boundary between the BCS solutions in both layers. If the FFLO state is the ground state in the FM layer ($q_f = g \neq 0$), the of equations (28) describes matching between the *ground* BCS state in the S layer and the *virtual* BCS state in the FM layer. The lower Eq. (28) corresponds to continuous matching on the FM/S boundary between the *real* FFLO solution in the FM layer and the *virtual* FFLO one in the S layer. Thus the boundary conditions (28) mean that the continuous matching takes place only between the pair amplitudes of identical symmetry.

In reality, of course, the pair amplitudes $F_s(\boldsymbol{\rho},+0,\omega)$ and $F_f(\boldsymbol{\rho},-0,\omega)$ are matching continuously via transitional (or crossover) kernels K_{sf} and K_{fs}. This fact can be easily seen, if we rewrite the Eq. (6) for the S and FM regions separately. Thus the exponential term $\exp(\pm i\mathbf{g}\boldsymbol{\rho})$ in Eqs. (28) is simply the remainder (trace) of a corresponding transitional kernel. If we take into account that the Gor'kov functions $F_s(\boldsymbol{\rho},+0,\omega)$ and $F_f(\boldsymbol{\rho},-0,\omega)$ depend on $\boldsymbol{\rho}$ according to the Eqs. (13), we can see that the Eq. (28) is satisfied identically. This means that integral boundary value problem (6) admits the solutions (13) with the different 2D wave vectors $\mathbf{q}_f$ and $\mathbf{q}_s$ with possible mismatch $\mathbf{g} = \mathbf{q}_f - \mathbf{q}_s$. In particular, Eq. (2) admits a competition between different types of the order parameter symmetry, say, between the homogeneous BCS pairing with $q_s = 0$ and the oscillating FFLO pairing with $q_f \neq 0$ in the framework of common *s*-type pairing scenario.

In case of the dirty limit, when the electron mean free path $l_{s(f)} = v_{s(f)}\tau_{s(f)}$ becomes much smaller than coherence length $\xi_{s(f)} = v_{s(f)}/2\pi T$ and spin stiffness length $a_f = v_f/2I$, the boundary value problem (24), (25) for the Eilenberger function $\Phi(\mathbf{p},\mathbf{q},z,\omega)$ is reduced to the corresponding one for the Gor'kov function $F(\mathbf{q},z,\omega)$. This dirty limit boundary value problem (see [3,13]) in the coordinate space includes the Usadel-like differential equations

$$\left[\omega + iI(z) - \frac{D(z)}{2}\left(\nabla_\rho^{\,2} + \frac{\partial^2}{\partial z^2}\right)\right] F(\boldsymbol{\rho},z,\omega) = \Delta(\boldsymbol{\rho},z) \tag{29}$$

and boundary conditions which connect the flux of $F(\boldsymbol{\rho},z,\omega)$ with its jump at the FMS interface

$$\begin{aligned} \frac{4D_s}{\sigma_s v_s}\frac{\partial F_s(\boldsymbol{\rho},z,\omega)}{\partial z}\bigg|_{z=+0} &= F_s(\boldsymbol{\rho},+0,\omega) - F_f(\boldsymbol{\rho},-0,\omega)e^{-i\mathbf{g}\boldsymbol{\rho}}; \\ D_s N_s \frac{\partial F_s(\boldsymbol{\rho},z,\omega)}{\partial z}\bigg|_{z=+0} &= D_f N_f \frac{\partial F_f(\boldsymbol{\rho},z,\omega)e^{-i\mathbf{g}\boldsymbol{\rho}}}{\partial z}\bigg|_{z=-0}. \end{aligned} \tag{30}$$

Here $D_{s(f)}$ are the diffusion coefficients and $\sigma_{s(f)}$ are the transmission coefficients which are related by the detailed balance conditions $\sigma_s v_s N_s = \sigma_f v_f N_f$ [3,8]. These coefficients $\sigma_{s(f)}$ are defined via average over the angle between the direction of electron velocity and a normal to the FM/S interface, i.e.

$$\sigma_{s(f)} = \left\langle \frac{\sigma v_{s(f)z}}{(1-\sigma) v_{s(f)}} \right\rangle .$$

Many possible solutions of the dirty limit boundary value problem (29), (30) have been well studied and can be found in Refs. [3,4,8,13]. It was established that *nonmonotonous* dependence of the FM/S multilayers T_c on the FM layer thickness d_f is caused by oscillations of the Cooper pairs flux (30) through the S/FM interface due to the 3D-1D-3D phase transitions cascade via *Umklapp processes*. Therefore below we shall below consider the simple examples of the BCS and FFLO pairing mutual accommodation in the *clean* FM/S nanostructures.

5. Proximity Effect between Semiinfinite FM and S Metals

As one of applications we will solve the boundary value problem (24), (25) for the contact between two pure ($\tau_s^{-1} = \tau_f^{-1} = 0$) semi-infinite metals FM ($-\infty < z < 0$) and S ($0 < z < \infty$). For this purpose we rewrite the self-consistency equation (1) and differential equation (24) separately for the layers FM and S. For the S layer we get

$$\Delta_s(\mathbf{q}_s, z) = 2\lambda_s \pi T \,\mathrm{Re} \sum_{\omega > 0} \int \frac{d\Omega_p}{4\pi} \Phi_s(\mathbf{p}, \mathbf{q}_s, z, \omega) ; \qquad (31)$$

$$\left[2\tilde{\omega}_s - v_{sz}\xi_{sz} \frac{\partial^2}{\partial z^2} \right] \Phi_s(\mathbf{p}, \mathbf{q}_s, z, \omega) = 2\Delta_s(\mathbf{q}_s, z), \quad 0 < z < \infty , \qquad (32)$$

where the following notations are used

$$\xi_{sz} = \frac{v_{sz}}{2\tilde{\omega}_s}; \quad 2\tilde{\omega}_s = 2\omega + i\left(\mathbf{v}_{s\perp}\mathbf{q}_s\right). \qquad (33)$$

Similarly we get for the FM layer

$$\Delta_f(\mathbf{q}_f, z) = 2\lambda_f \pi T \,\mathrm{Re} \sum_{\omega > 0} \int \frac{d\Omega_p}{4\pi} \Phi_f(\mathbf{p}, \mathbf{q}_f, z, \omega) ; \qquad (34)$$

$$\left[2\tilde{\omega}_f - v_{fz}\xi_{fz} \frac{\partial^2}{\partial z^2} \right] \Phi_f(\mathbf{p}, \mathbf{q}_f, z, \omega) = 2\Delta_f(\mathbf{q}_f, z), \quad -\infty < z < 0 ; \qquad (35)$$

$$\xi_{fz} = \frac{v_{fz}}{2\tilde{\omega}_f}; \quad 2\tilde{\omega}_f = 2\omega + i\left[2I + (\mathbf{v}_{f\perp} \cdot \mathbf{q}_f) \right]. \qquad (36)$$

The boundary conditions corresponded to Eqs. (32), (35) take the form

$$\xi_{sz}\frac{\partial\Phi_s(\mathbf{p},\mathbf{q}_s,z,\omega)}{\partial z}\Big|_{z=+0}=\xi_{fz}\frac{\partial\Phi_f(\mathbf{p},\mathbf{q}_f,z,\omega)}{\partial z}\Big|_{z=-0}=$$
$$=\frac{\sigma}{2(1-\sigma)}\left[\Phi(\mathbf{p},\mathbf{q}_s,+0,\omega)-\Phi(\mathbf{p},\mathbf{q}_f,-0,\omega)\right]. \tag{37}$$

Unlike all previous papers on the FM/S structures (see the reviews [3,4] and references therein), where only the assumption $\lambda_f = 0$ is considered, we take into account the own electron-electron interaction in the FM layer. Thus, we admit the presence of the cause for the origin of its own superconductivity in ferromagnet ($\lambda_f \neq 0$). On the other hand, due to the exchange field $I \neq 0$ presence in the FM layer not only the BCS but also the FFLO type of pairing is possible. Because of this, the question about real matching between the oscillating (FFLO) and the constant sign (BCS) pair amplitudes at the FM/S boundary becomes very interesting.

We will search for the solutions of Eqs. (32), (35) in the form

$$\Phi_s(\mathbf{p},\mathbf{q}_s,z,\omega)=\frac{\Delta_{s0}}{\tilde{\omega}_s}+\frac{A}{\tilde{\omega}_s}\exp(-\frac{z}{\xi_{sz}});\quad 0<z<\infty, \tag{38}$$

$$\Phi_f(\mathbf{p},\mathbf{q},z,\omega)=\frac{\Delta_{f0}}{\tilde{\omega}_f}+\frac{B}{\tilde{\omega}_f}\exp(\frac{z}{\xi_{fz}});\quad -\infty<z<0. \tag{39}$$

Here Δ_{s0} and Δ_{f0} are the undisturbed values of the order parameter, which are determined deep in the interior of the S and FM layers, correspondingly, i.e. at $|z| >> \xi_{sz}, \xi_{fz}$. The unknown constants A and B must be defined by substitution of the solutions (38) and (39) into the boundary conditions (37). The resulting Eilenberger functions take the form

$$\Phi_s(\mathbf{p},\mathbf{q}_s,z,\omega)=\frac{\Delta_{s0}}{\tilde{\omega}_s}-\frac{\sigma}{2}\left[\frac{\Delta_{s0}}{\tilde{\omega}_s}-\frac{\Delta_{f0}}{\tilde{\omega}_f}\right]\exp(-\frac{z}{\xi_{sz}});\quad 0<z<\infty,$$
$$\Phi_f(\mathbf{p},\mathbf{q}_f,z,\omega)=\frac{\Delta_{f0}}{\tilde{\omega}_f}+\frac{\sigma}{2}\left[\frac{\Delta_{s0}}{\tilde{\omega}_s}-\frac{\Delta_{f0}}{\tilde{\omega}_f}\right]\exp(\frac{z}{\xi_{fz}});\quad -\infty<z<0. \tag{40}$$

It is easy to check that in the ideal transparency case ($\sigma = 1$) at the boundary $z = 0$ both functions coincide each other, i.e.

$$\Phi_s(\mathbf{p},\mathbf{q}_s,+0,\omega)=\Phi_f(\mathbf{p},\mathbf{q}_f,-0,\omega)=\frac{1}{2}\left[\frac{\Delta_{s0}}{\tilde{\omega}_s}+\frac{\Delta_{f0}}{\tilde{\omega}_f}\right]. \tag{41}$$

If $\sigma < 1$, then the Eilenberger function undergoes the jump at the FM/S boundary, and this skip defines its flux (37). As it follows from the Eqs. (40) the scale of the proximity effect region is of order of ($\xi_{sz} + \xi_{fz}$). In this interface layer the periodical *surface states* are located with the mixed BCS+FFLO type of pairing.

It can be shown that in the S region a 2D vector of the pairs $\mathbf{q}_s$ must be exactly equal to zero. Actually, deep within the S region when $z >> \xi_{sz}$ the Eilenberger function tends to a constant, i.e.

$$\Phi_s(\mathbf{p},\mathbf{q}_s,z,\omega) \cong \frac{\Delta_{s0}}{\tilde{\omega}_s} = \frac{2\Delta_{s0}}{2\omega + i\left(\mathbf{v}_{s\perp}\mathbf{q}_s\right)}. \tag{42}$$

Substituting Φ_s (42) into the self-consistency equation (31) and summing over ω, we get the following expression for the reduced critical temperature $t_s = T_c/T_{cs}$ deep in the interior of S region

$$\ln t_s = \Psi\left(\frac{1}{2}\right) - \frac{1}{2}\mathrm{Re}\int_{-1}^{1}\Psi\left(\frac{1}{2} + \frac{iq_s v_s x}{4\pi T_{cs} t}\right)dx. \tag{43}$$

Here T_{cs} is the critical temperature of the isolated bulk superconductor S defined by the constant of electron-electron interaction λ_s. Then the T_c maximum condition leads strictly to condition that $\mathbf{q}_s = 0$ and $T_c = T_{cs}$ deep inside the S region. As it follows from the Eqs. (13), (31) and (40) the order parameter Δ_s and the functions F_s and Φ_s in the S region are nearly monotone constant sign functions of the variable z and they do not oscillate in the FM/S boundary plane. The exception is the layer $0 < z < \xi_{sz}$ near the FM/S boundary plane, where the oscillations in $\boldsymbol{\rho}$ are imported from the FM layer, their amplitude damps with rise of z.

In turn deep inside the FM region, where $|z| >> \xi_{fz}$, the Eilenberger function is also nearly constant, i.e.

$$\Phi_f(\mathbf{p},\mathbf{q}_f,z,\omega) \cong \frac{\Delta_{f0}}{\tilde{\omega}_f} = \frac{2\Delta_{f0}}{2\omega + i\left[2I + (\mathbf{v}_{f\perp}\mathbf{q}_f)\right]}. \tag{44}$$

The corresponding expression for the $t_f = T_c/T_{cf}$ deep inside the FM region has the form

$$\ln t_f = \Psi\left(\frac{1}{2}\right) - \frac{1}{2}\mathrm{Re}\int_{-1}^{1}\Psi\left(\frac{1}{2} + \frac{i[2I + q_f v_f x]}{4\pi T_{cf} t}\right)dx, \tag{45}$$

where T_{cf} is the critical temperature of the ferromagnetic superconductor FM in the assumption of absence of the exchange field I. The T_{cf} value is defined by only the electron-electron interaction constant λ_f, namely

$$\frac{1}{\lambda_f} = \ln\left(\frac{2\gamma\omega_D}{\pi T_{cf}}\right).$$

The equation (45) describes the paramagnetic effect of the exchange field I in the BCS state with $q_f = 0$ and its partial compensation due to the FFLO state when the $q_f \neq 0$ arises (see discussion after Eq. (12) in Sec. 3). In case of the FFLO state realization in the FM region the order parameters Δ_f, F_f and Φ_f are the functions oscillating in $\mathbf{\rho}$ with period q_f^{-1}. These oscillations occur near the FM/S boundary in the layer $-\xi_{sz} < z < 0$ on the background of monotonically decreasing amplitude of the constant sign for the BCS correlations which penetrate from the S region (see formulae (40)).

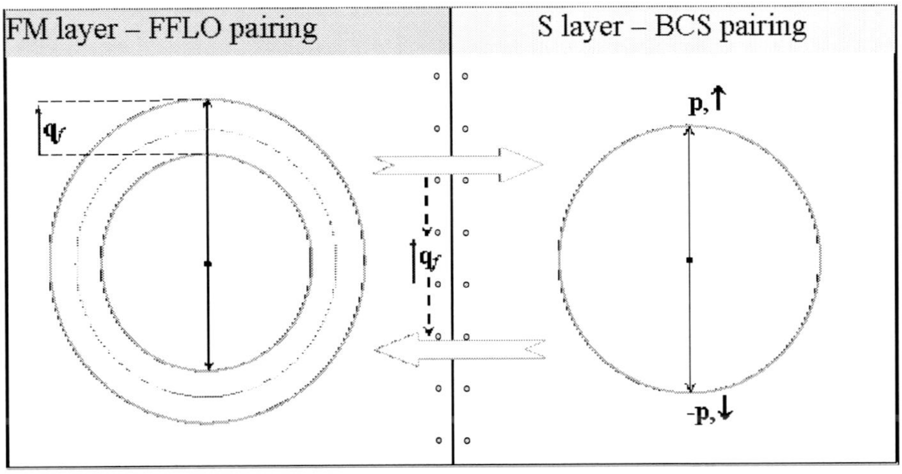

Figure 3. The processes of mutual transformation and Umklapp processes at the FM/S boundary. The Fermi surfaces are shown in both sides of contact. In the S side we have the BCS pairing with zero total momentum. In the FM side we have exchange splitted Fermi surface, and the FFLO pair with nonzero coherent momentum $\mathbf{q}_f$ is schematically depicted. Only one layer of damping FFLO lattice (small circles) is shown on each side of the FM/S interface. The thick horizontal arrows denote the superconducting pairs leaving the FM(S) region for the S(FM) one, respectively. The dashed vertical lines symbolize the transfer of excess momentum $\mathbf{q}_f$ from the FFLO pair to the BCS pair via the FFLO lattice (the Umklapp process) and, consequently, the transformation of the last one to the new FFLO pair.

Near the FM/S boundary in the layer with thickness $\sim (\xi_{sz} + \xi_{fz})$, where the 2D periodical structure (lattice) of surface states with the mixed BCS+FFLO mechanism of pairing is realized, the critical temperature T_c varies as function of z. Its middle value we can approximately estimate for the case of ideal transparency ($\sigma = 1$) with help the Eq. (41) for the Eilenberger function at the surface $z = 0$. In the simplest case of the FM and S metals with identical electronic parameters, except that λ_f, λ_s and I are different, we get near boundary FM/S for the middle critical temperature $t(0) = T_c(0)/T_{cs}$

$$\ln t(0) \approx \frac{\lambda_f - \lambda_s}{\lambda_s(\lambda_s + \lambda_f)} + \frac{\lambda_f}{\lambda_s + \lambda_f}\left[\Psi\left(\frac{1}{2}\right) - \frac{1}{2}\mathrm{Re}\int_{-1}^{1}\Psi\left(\frac{1}{2} + \frac{i[2I + q_f v_f x]}{4\pi T_{cs} t(0)}\right)dx\right]. \quad (46)$$

For example, if $\lambda_s > \lambda_f > 0$, the critical temperature increases as the function of z in going from the FM region to the S region, that is, from $T_{cf}(I)$ defined by the Eq. (45) up to T_{cs} deep inside the S region.

In case when the FFLO state in the FM region and the BCS one in the S region with $T_{cs} \sim T_{cf}(I)$ originate, the presence of transitional region ($\xi_{sz} + \xi_{fz}$) with surface states leads to mutual transformations between the FFLO and BCS pairs, which are realized via the *Umklapp processes*. Schematically these processes are shown in the Fig. 3. The FFLO pair leaving the FM region for the S layer gives its excess 2D momentum $\mathbf{q_f}$ to the 2D surface states lattice and transforms to the BCS pair in the S region . In turn the BCS pair passing from the S region into the FM region gains additional momentum $\mathbf{q_f}$ from the 2D lattice of surface states and transforms to the FFLO pair in the FM region .

6. Proximity Effect between thin FFLO and BCS Superconductors

Let us consider the planar contact between the pure FFLO superconductor FM, which occupies the region $-d_f < z < 0$, and the pure BCS superconductor S, which occupies the region $0 < z < d_s$. We will search for the solutions of boundary value problem (32), (35), (37) in the form which excludes the fluxes through the outer boundaries of the FM/S contact, i.e.

$$\Phi_s(\mathbf{p},\mathbf{q}_s,z,\omega) = \frac{\Delta_{s0}}{\tilde{\omega}_s} + A\frac{\cosh[(z-d_s)/\xi_{sz}]}{\cosh(d_s/\xi_{sz})}, \quad 0 < z < d_s;$$

$$\Phi_f(\mathbf{p},\mathbf{q}_f,z,\omega) = \frac{\Delta_{f0}}{\tilde{\omega}_f} + B\frac{\cosh[(z+d_f)/\xi_{fz}]}{\cosh(d_f/\xi_{fz})}, \quad -d_f < z < 0; \tag{47}$$

$$\xi_{sz} = \frac{v_{sz}}{2\tilde{\omega}_s}; \; \tilde{\omega}_s = \omega; \; \xi_{fz} = \frac{v_{fz}}{2\tilde{\omega}_f}; \; 2\tilde{\omega}_f = 2\omega + i(2I + \mathbf{v}_{\mathbf{f}\perp}\mathbf{q}_\mathbf{f}).$$

Here Δ_{s0} and Δ_{f0} are the initial values of the order parameter in the thin FM and S films. It is easy to check that these solutions (47) obey Eqs. (32). For simplicity we consider only the case of ideal transparency ($\sigma \to 1$). Substitution of solutions (47) in the boundary conditions (37) gives a set of algebraic equations on parameters A and B. The mutual influence of the FM and S metals is especially significant in the Cooper limit, where the layer thicknesses are small: $d_{f(s)} \ll (\xi_{f(s)}, a_f)$. In this case the order parameter and the Eilenberger function $\Phi(\mathbf{p},\boldsymbol{\rho},z,\omega)$ are practically constant along the z axis inside the FM and S layers, and the spatial variations are possible only in $\boldsymbol{\rho}$ (along the FM/S boundary plane). After simple calculations we find that the Eilenberger functions in the thin FM/S contact look like

$$\Phi_s(\mathbf{p},z,\omega) = \Phi_f(\mathbf{p},\mathbf{q}_f,z,\omega) = \frac{c_s\Delta_{s0} + c_f\Delta_{f0}}{\omega + ic_f(I + \mathbf{q}_f\mathbf{v}_{f\perp}/2)}; \quad c_{f(s)} = \frac{d_{f(s)}}{d_f + d_s}, \tag{48}$$

where c_f and c_s are the relative weights of the FM and S layers, respectively. At derivation of Eq. (48) for simplicity we considered the case when the contacting metals (FM and S) posses identical electronic parameters, except the fact that interelectron interactions $\lambda_{s(f)}$ and exchange splitting of the Fermi level can be different.

Summing over the frequency ω in the Gor'kov self-consistency Eqs. (31), (34), the following expression for the reduced superconducting transition temperature $t = T_c/T_{cs}$ of the FM/S bilayer can be derived

$$\ln t = -\frac{c_f(\lambda_s - \lambda_f)}{\lambda_s(c_s\lambda_s + c_f\lambda_f)} + \Psi\left(\frac{1}{2}\right) - \frac{1}{2}\mathrm{Re}\int_{-1}^{1} dx \Psi\left(\frac{1}{2} + \frac{i(2c_f I + xqv_\perp)}{4\pi T_{cs} t}\right), \quad (49)$$

where T_{cs} is the critical temperature of isolated S layer, and x is the cosine of angle between vectors $\mathbf{q}_f = \mathbf{q}_s = \mathbf{q}$ and $\mathbf{v}_{f\perp} = \mathbf{v}_{s\perp} = \mathbf{v}_\perp$. In the right side of Eq. (49) the first term describes a decrease of T_c since usually $\lambda_s > \lambda_f$. It corresponds to the de Gennes result [14] for the usual normal metal/superconductor bilayer. The second and third terms describe the T_c suppression due to paramagnetic effect of exchange field I and its partial compensation due to a creation of the FFLO state with $q \neq 0$.

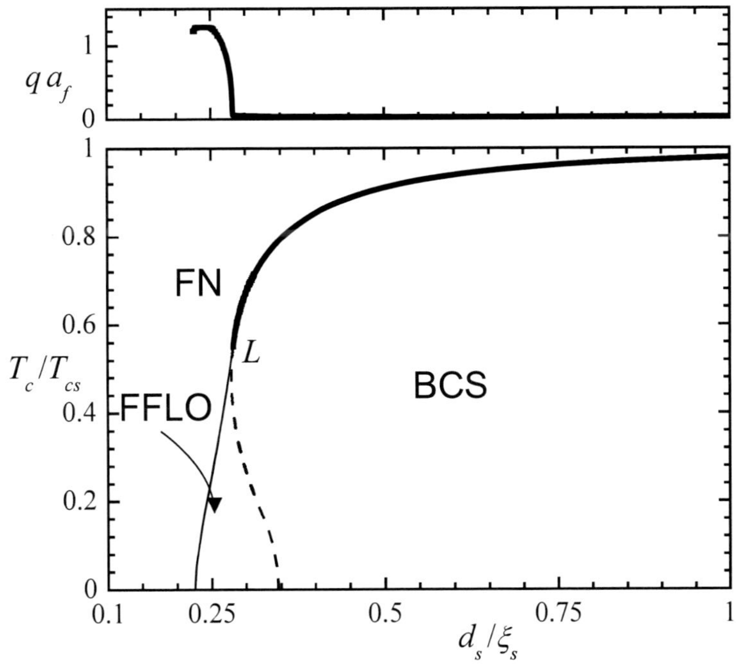

Figure 4. Phase diagrams $T_c(d_s)$ (lower panel) and $q(d_f)$ (upper panel) of the clean FM/S bilayer. The thick solid curve is the second order phase transitions curve, which separates the BCS state ($q = 0$) and ferromagnetic normal state (FN). The thin solid line below the Lifshitz point L corresponds to the transition of the second order between the FFLO state with $q_f \neq 0$ and the FN state. The thin dashed curve after the Lifshitz point L represents the unreliazable virtual BCS state with $q = 0$. We used the following parameters in numerical calculations: $d_f/a_f = 0.1$; $\xi_s/a_f = 10$. For brevity the line of first order transitions (see Fig. 2) is not shown.

The $T_c(d_s)$ dependence for such an FM/S bilayer is depicted in Fig. 3 for case $\lambda_s = \lambda_f$. The concrete value of q is found from the T_c maximum condition for each value of d_s (see upper panel of Fig. 3). It is easy to see that there are two types of superconducting states in the whole FM/S bilayer. First of them is the BCS state with $q = 0$ and constant order parameter $\Delta_{f(s)}$ is realized at large d_s. With d_s decrease in the Lifshitz point L [12] this BCS state changes to the FFLO one, in which $q_f \neq 0$ and both order parameters $\Delta_{f(s)}(\boldsymbol{\rho})$ and pair amplitudes $\Phi_{f(s)}(\mathbf{p},\boldsymbol{\rho},\omega)$ are oscillating functions in the FM/S boundary plane (x-y) with period of order of the spin stiffness length a_f.

It is not surprisingly that the $T_c(d_f)$ curve for considered FM/S system with thin layers and ideal transparency will look as the T_c dependence depicted in Fig.2 for homogeneous ferromagnetic superconductor. Actually, due to strong proximity effect the effective average of all interactions takes place. To see the $T_c(d_f)$ and $q(a_f)$ curves in Fig.2 (lower and upper panels, correspondingly) we should consider only the upper axes (d_f/a_f) for both panels and right axes (T_c/T_{cs} and qa_f) with corresponding labeled dimensionless values. The following parameters are used in calculations: $d_s/\xi_s = 0.25$; $\xi_s/a_f = 10$.

From analysis of Figs. 2 and 4 it follows that the clean FM/S bilayers could be used as ideal model systems for direct observation of the FFLO state realization.

In general case, when the metals FM and S possess by the different electronic Fermi surface parameters, the weight coefficients c_s and c_f acquire the angle dependence and the Eqs. (48), (49) become essentially more complicated.

Above we have considered the case, when 2D pairs momentums are equal each other, i.e. $\mathbf{q}_f = \mathbf{q}_s = \mathbf{q}$. However the obtained formula (49) and the T_c and q dependences on thicknesses d_f and d_s (see Figs. 2 and 4, correspondingly) still stand (with the replacing $q = c_f q_f$) in the case $\mathbf{q}_s = 0$ and $\mathbf{q}_f \neq 0$. This state being correspond to realization only the BCS state in the S layer, but in the FM layer except the BCS state at some thickness d_f the FFLO state could be arose. The continuous matching of pairs amplitudes in the latter case could be occur only via the excited (virtual) surface states FFLO and BCS as this described in the end of Sec. 4. In this case the mutual transformations between the FFLO and BCS pairs should be happen through the Umklapp processes. The answer on the question what from two possible states $\mathbf{q}_f = \mathbf{q}_s = \mathbf{q}$ or $\mathbf{q}_s = 0$, $\mathbf{q}_f \neq 0$ possess by the smallest free energy, can be get if we take into account the terms of higher order than Δ^2 in the Ginzburg-Landau expansion. But the investigation of this interesting question is beyond the scope of present work.

7. π Phase Superconductivity in FM/S/FM Trilayers

Here we consider the layer of pure BCS superconductor S, which occupies the region $0 < z < d_s$ and it is sandwiched between two pure FFLO superconductors FM and FM' occupying the regions $-d_f < z < 0$ and $d_s < z < d_s + d_f$, correspondingly. We will search for the solutions of boundary value problem (32), (35), (37) in the form which excludes the fluxes through the outer boundaries ($z = -d_f$ and $z = d_s + d_f$) of the FM/S/FM' trilayer, i.e.

$$\begin{aligned}
\Phi_s(\mathbf{p},\mathbf{q}_s,z,\omega) &= \frac{\Delta_{s0}}{\tilde{\omega}_s} + A\frac{\cosh[(z-d_s/2)/\xi_{sz}]}{\cosh(d_s/2\xi_{sz})} + C\frac{\sinh[(z-d_s/2)/\xi_{sz}]}{\sinh(d_s/2\xi_{sz})}, \quad 0<z<d_s; \\
\Phi_f(\mathbf{p},\mathbf{q}_f,z,\omega) &= \frac{\Delta_{f0}}{\tilde{\omega}_f} + B\frac{\cosh[(z+d_f)/\xi_{fz}]}{\cosh(d_f/\xi_{fz})}, \quad -d_f<z<0; \\
\Phi'_f(\mathbf{p},\mathbf{q}_f,z,\omega) &= \frac{\Delta_{f0}}{\tilde{\omega}'_f} + B'\frac{\cosh[(z-d_s-d_f)/\xi'_{fz}]}{\cosh(d_f/\xi'_{fz})}, \quad d_s<z<d_s+d_f,
\end{aligned} \tag{50}$$

where the correlation lengths ξ_{sz}, ξ_{fz} are defined by Eqs. (47) and ξ'_{fz} is defined by formula

$$\xi'_{fz} = \frac{v_{fz}}{2\tilde{\omega}'_f}; \; 2\tilde{\omega}'_f = 2\omega \pm i(2I + \mathbf{v}_{f\perp}\mathbf{q}_f). \tag{51}$$

Here the upper sign corresponds to parallel orientation of the FM and FM' layers magnetizations (0 phase magnetic state), whereas the lower sign corresponds to their antiparallel alignment (π phase magnetic state). The boundary conditions on the first FM/S interface ($z = 0$) are given by Eqs. (37). The boundary conditions on the second S/FM' interface ($z = d_s$) look like

$$\begin{aligned}
&\Phi_s(\mathbf{p},\mathbf{q}_s,d_s-0,\omega) = \pm\Phi'_f(\mathbf{p},\mathbf{q}_f,d_s+0,\omega); \\
&\xi_{sz}\left.\frac{\partial\Phi_s(\mathbf{p},\mathbf{q}_s,z,\omega)}{\partial z}\right|_{z=d_s-0} = \pm\xi'_{fz}\left.\frac{\partial\Phi'_f(\mathbf{p},\mathbf{q}_f,z,\omega)}{\partial z}\right|_{z=d_s+0},
\end{aligned} \tag{52}$$

where the upper sign responds for the usual 0 phase superconductivity with coinciding phases of the order parameters Δ_f and Δ'_f in the FM and FM' layers. Whereas the lower sign corresponds to the unconventional π phase superconductivity at which the sign of the order parameter Δ_f changes on the opposite one when the S interlayer crossing. Up to now this π phase superconductivity state for the FM/S/FM trilayers was considered as impossible in principle [3,4]. The reason is that in the previous theories (see Refs. in [3, 4]) the electron-electron interaction in the FM layer was neglected, i.e. these theories were restricted by the partial case $\lambda_f=0$ only.

Thus, unlike the FM/S bilayers, in the FM/S/FM' trilayers case the four different states ($\varphi\chi$) are possible. These states are distinguished by phase φ of the superconducting (Δ_f) and phase χ of magnetic (I) order parameters in the neighboring FM layers. In the both π-magnetic states (0π and ππ) the neighboring FM layers magnetizations are antiparallel to one another and their paramagnetic effect is mutually compensated not only in the S interlayer but also in these FM layers interior due to ideal transparency and the Cooper limit. Actually our calculations for the simplest case of metals FM and S with identical electronic structure show that the Eilenberger functions in the 0π and ππ states look like

$$\begin{aligned}
&\Phi_s^{0\pi}(\mathbf{p},\omega)=\Phi_f^{0\pi}(\mathbf{p},\mathbf{q}_f,\omega)=\Phi'^{0\pi}_f(\mathbf{p},\mathbf{q}_f,\omega)=\frac{c_s\Delta_{s0}+c_f\Delta_{f0}}{\omega};\ c_f+c_s=1;\\
&\Phi_s^{\pi\pi}(\mathbf{p},\omega)=\Phi_f^{\pi\pi}(\mathbf{p},\mathbf{q}_f,\omega)=-\Phi'^{\pi\pi}_f(\mathbf{p},\mathbf{q}_f,\omega)=\frac{c_s\Delta_{s0}}{\omega};\ c_f=\frac{2d_f}{2d_f+d_s},
\end{aligned} \tag{53}$$

where c_f and c_s are the relative weights of the FM and S layers, respectively, in the FM/S/FM' trilayer. The pair-breaking factors of the neighbor FM and FM' layers are complex conjugated and mutually compensate each other. Therefore the T_c of such 0π and ππ trilayers depends only upon the sign and magnitude of the electron-electron interaction λ_f in the FM layers. Indeed we get the following equations for the reduced critical temperatures $t^{0\pi}=T_c^{0\pi}/T_{cs}$ and $t^{\pi\pi}=T_c^{\pi\pi}/T_{cs}$

$$\ln t^{0\pi}=-\frac{c_f(\lambda_s-\lambda_f)}{\lambda_s(c_s\lambda_s+c_f\lambda_f)},\quad \lambda_f>0;\quad \ln t^{\pi\pi}=-\frac{c_f}{c_s\lambda_s},\quad \lambda_f<0. \tag{54}$$

It is interesting to note, that T_c of the ππ state due to π phase superconductivity realization does not depend on the value of $|\lambda_f|$. It is easy to see, the critical temperature $T_c^{\pi\pi}$ is higher than $T_c^{0\pi}$ one, if $\lambda_f<0$. In the case $\lambda_f=0$ these 0π and ππ states coincide with each other. The possibility of two 0π and ππ states which are different in superconductivity is very surprising fact for the FM/S/FM trilayers since it is usually considered that the π phase superconductivity in such systems is principally impossible. On the one side, existence of the π phase superconductivity in the FM/S/FM trilayers means that coupling between the neighboring FM layers is provided by the superconducting correlations, i.e. the role of true order parameter is played by the pair amplitude $F(\rho,z,\omega)$ rather than $\Delta(\rho,z)$. On the other hand, it is known [3,4], that the states of the superlattice include the states of the trilayers due to symmetry reasons. Therefore we can qualitatively analyze the experimental data by Goff *et al.* [5]. Indeed, the existence of such π magnetic states (53), (54) allows us to explain the unexpected weak depression of superconductivity which has been experimentally found in the short period Gd/La superlattices. The measured T_c of this Gd/La superlattice was 5 K that practically coincides with critical temperature of bulk lanthanum sample! This means that in the Gd/La superlattice the 0π type state is realized rather than ππ one and $\lambda_f\approx\lambda_s$ since $T_c^{0\pi}\approx T_{cs}$ as it follows from the first of Eqs. (54). That is, the electron-electron interaction in the isolated Gd film corresponds to attraction but its own superconductivity is depressed by the strong exchange field $I\gg T_c$. However, due to compensation of the exchange fields in the 0π state of the Gd/La superlattice the superconductivity of the Gd film is restored again.

The order parameter $\Delta_f(\mathbf{q}_f,z)$ and $\Delta_s(z)$ profiles corresponding to the 0π and ππ states in the FM/S/FM system with thin layers are schematically depicted in the Figs. 5 and 6. The physical meaning of qualitatively different behavior of the superconducting order parameter in these 0π and ππ states becomes clear, if we rewrite the Eqs. (54) for the critical temperatures $T_c^{0\pi}$ and $T_c^{\pi\pi}$ in the form

$$\ln\frac{2\gamma\omega_D}{\pi T_c^{0\pi}}=\frac{1}{c_s\lambda_s+c_f\lambda_f},\quad \lambda_f>0;$$
$$\ln\frac{2\gamma\omega_D}{\pi T_c^{\pi\pi}}=\frac{1}{c_s\lambda_s},\quad \lambda_f<0, \tag{55}$$

where γ =1.781 is the Euler constant. We can see that in the case of 0-phase superconducting state with $\lambda_f > 0$, the $T_c^{0\pi}$ is defined by the *effective* electron-electron attraction $\lambda_{eff}^{0\pi} = c_s\lambda_s + c_f\lambda_f$ averaged over the FM/S/FM trilayer as this is shown in Fig. 5. Whereas in the case of π-phase superconducting state with $\lambda_f < 0$, the $T_c^{\pi\pi}$ is defined by the electron-electron attraction $\lambda_{eff}^{\pi\pi} = c_s\lambda_s$ of the S layer only, since the contributions of the positive and negative areas under the -Δ_f and Δ_f lines in Fig. 6 are mutually compensate each other.

It is necessary to note that, in the case of electron-electron repulsion ($\lambda_f < 0$) in the FM layers, the $T_c^{\pi\pi} > T_c^{0\pi}$ and the π-phase superconducting state (ππ) is energetically favorable in comparison with 0-phase superconducting state (0π). In the opposite case with electron-electron attraction ($\lambda_f > 0$) in the FM layers, on the contrary the $T_c^{0\pi}$ will be greater than $T_c^{\pi\pi}$.

Moreover, in the FM/S/FM trilayers there are exist two other states 00 and π0 in which the neighbor FM layers magnetizations are parallel to one another. Their critical temperatures t^{00} and $t^{\pi 0}$ are as follows

$$\ln t^{00}=\ln t^{0\pi}+\Psi\left(\frac{1}{2}\right)-\frac{1}{2}\mathrm{Re}\int_{-1}^{1}dx\Psi\left(\frac{1}{2}+\frac{ic_f(2I+xq_fv_f)}{4\pi T_{cs}t}\right);$$
$$\ln t^{\pi 0}=\ln t^{\pi\pi}+\Psi\left(\frac{1}{2}\right)-\frac{1}{2}\mathrm{Re}\int_{-1}^{1}dx\Psi\left(\frac{1}{2}+\frac{ic_f(2I+xq_fv_f)}{4\pi T_{cs}t}\right). \tag{56}$$

These 00 and π0 states possess intensified paramagnetic effect of exchange field I, and being in close analogy to the FM/S bilayer case, admit competition between the smooth BCS and oscillating FFLO states with the FM layers thickness d_f increasing. These 00 and π0 states have lower T_c in relation to the previous 0π and ππ ones. Probably they could be observed in the presence of the external magnetic field $H > H_{coer}$ (where H_{coer} is the coercive field of the FM film) if, of course, the averaged exchange field c_fI is not too strong. By the way, in the field cooling Gd/La superlattice with parallel orientations of magnetizations in all the Gd layers the superconducting transition has been completely suppressed [5] in accordance with our theory. This means, evidently, that thicknesses of the Gd and La layers were not optimal in the experiment [5]. Hereafter, for more precise comparison with experiments it is also necessary to extend equations (29),(34),(35) to the finite transparency of the FM/S boundary and to the arbitrary thicknesses and electronic parameters of the FM and S layers. Let's note that analogous to Eqs. (28)-(35) results for the FM/S bilayers and FM/S/FM trilayers in the Cooper limit can be obtained also in the dirty limit with use of the Usadel like boundary value problem (25), (26).

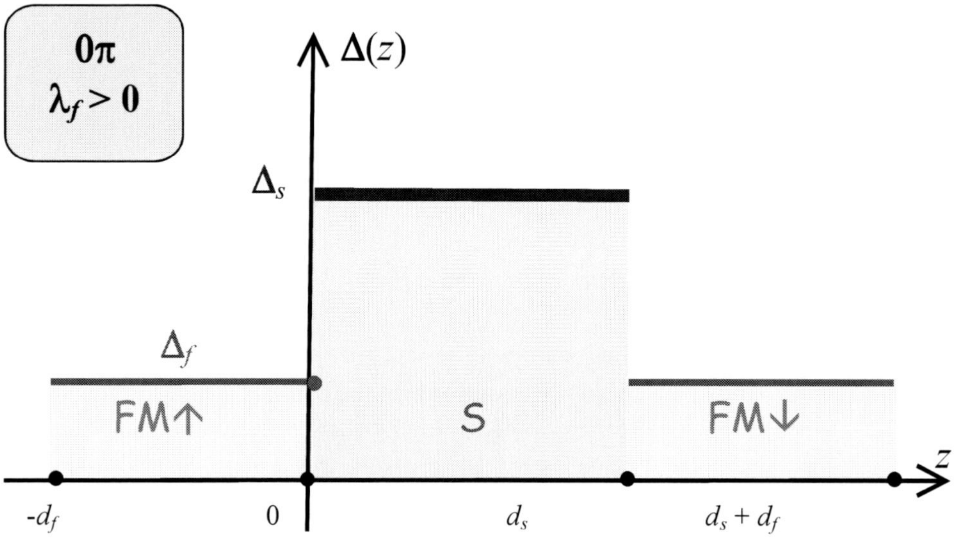

Figure 5. The schematically pictured superconducting order parameters $\Delta_f(\mathbf{q}_f,z)$ and $\Delta_s(z)$ profiles corresponding to the 0π state in the FM/S/FM trilayer.

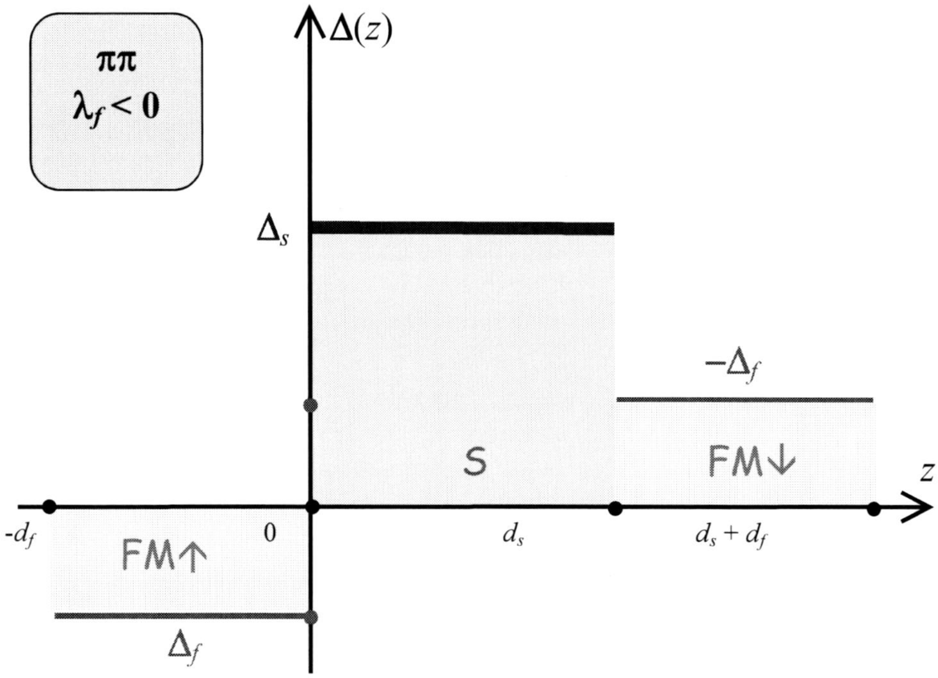

Figure 6. The schematically pictured superconducting order parameters $\Delta_f(\mathbf{q}_f,z)$ and $\Delta_s(z)$ profiles corresponding to the ππ state in the FM/S/FM trilayer.

8. Conclusion

Thus, for the FM/S nanostructures we have derived a new 3D quasiclassical boundary value problem, which includes differential equation (24) for the Eilenberger function $\Phi_{f(s)}(\mathbf{p},\mathbf{q},z,\omega)$ and corresponding boundary conditions (25) that relate the flux of $\Phi_{f(s)}(\mathbf{p},\mathbf{q},z,\omega)$ with its jump at the FM/S interface. This boundary value problem allows competition between the unconventional FFLO and the usual BCS types of pairing in the FM/S structures.

On this base we investigate the proximity effect between the contacting bulk pure metals FM and S. If both metals are the potential superconductors with $\lambda_s > \lambda_f > 0$, the critical temperature grows as the z function from $T_c(I)$ deep inside the FM layer up to T_{cs} deep inside the S layer. The constant sign $\Delta_s(z)$ BCS state is only realized deep inside the S layer. Whereas deep inside the FM layer with suitable parameters not only the BCS state ($q_f = 0$) but also FFLO one ($q_f \neq 0$) can be realized. This oscillating (along FM/S boundary) FFLO solution can not immediately match with the smooth BCS one at the FM/S interface. That is why a continuous matching between BCS state in the S layer and the FFLO state in the FM layer is possible only via the transitional region ($\xi_{sz} + \xi_{fz}$) with damped periodical lattice of the surface states. In this case, the processes of mutual transformation between the FFLO and BCS pairs occur only via Umklapp processes involving the virtual BCS state unrealized in the FM layer. During this Umklapp process the FFLO pair, leaving the FM layer for the S layer, gives up its excess 2D momentum $\mathbf{q}_f$ to the 2D surface states lattice (with mixed BCS+FFLO pairing) and transforms to the BCS pair in the S region. In turn, the BCS pair, passing from the S layer to the FM one, acquires this missing momentum $\mathbf{q}_f$ from the 2D lattice of surface states and transforms to the FFLO pair in the FM region.

We have also investigated the phase diagrams $T_c(d_f)$ and $T_c(d_s)$ for the thin FM/S bilayer in the pure Cooper limit when the thicknesses of the FM layer d_f and the S layer d_s are smaller than the corresponding coherence lengths. It was found that in this case two scenarios of the FFLO state realization in the FM/S bilayer are possible. First of them is the following. In the case of ideal transparency the effective average of all interactions and the exchange field I on whole FM/S specimen takes place due to strong proximity effect. In this case 2D wave vectors of the pairs in both FM and S layers are equal one to another, i.e. $q_f = q_s = q$. Optimal value of q is searched for from the $T_c(d_f)$ and $T_c(d_s)$ maximum condition. However another scenario is also possible. Our formula for T_c remains valid in the case $q_s = 0$ and $q_f \neq 0$ with the replacing $q \rightarrow c_f q_f$, where c_f is the volume fraction of the FM layer. This state corresponds to a realization only the BCS state in the S layer, but in the FM layer either the BCS state or the FFLO state can arise. The last choice depends on the relation between thicknesses d_f and d_s. At the small FM layer thickness d_f (large S layer thickness d_s) the *ground state* in both layers is the BCS one and the FFLO state with smaller T_c is the *virtual one*. At the same time for the large d_f (small d_s) the FFLO state with oscillating along the FM/S boundary pair amplitude becomes the ground state in the FM layer. In the latter case the continuous matching of pair amplitudes can be occurred only via the excited (virtual) surface states FFLO (in the S layer) and BCS (in the FM layer), and the mutual transformations between the FFLO and BCS pairs should go through the Umklapp processes. The answer on the question what from two possible states $q_f = q_s = q$ or $q_s = 0$ and $q_f \neq 0$ possess by the smallest free energy can be get if we take into account the terms of higher order than Δ^2 in the Ginzburg-

Landau expansion. But the investigation of this interesting question is beyond the scope of present work.

For the FM/S/FM trilayers we have found the novel π0 and ππ states with π phase superconductivity at repulsive interaction between electrons in the FM layers. The possibility of the π phase superconductivity in this trilayers is also a new and interesting fact for studying by the spin polarized transport and Josephson effect technique [3,4]. In the FM/S superlattices there are 0π and ππ states with compensation of strong exchange field I due to antiferromagnetic orientation of the neighboring FM layers magnetizations. This fact allows us to explain a surprisingly high $T_c \sim 5$K of the short period Gd/La superlattice and to predict the magnitude and sign of electron-electron interaction in the gadolinium metal.

Thus on the base of our investigations of the FM/S bilayers and FM/S/FM trilayers we can propose the *method of the proximity effect spectroscopy*, which makes possible the detection of unknown electronic parameters of contacting metals: for example, one can determine the exchange field I and the electronic correlations parameter λ_f in ferromagnet FM, if well known BCS superconductor S is used as a probe. The extension of our results on the high temperature superconductors (HTS) and the S/HTS and FM/HTS study could be give good method for a probe of the electronic correlations type and the order parameter symmetry in HTS material.

Acknowledgments

The authors (MGKh and YuNP) are grateful to the Max Planck Institute for the Physics of Complex Systems (Dresden, Germany) for hospitality and to Professor P. Fulde and the participants of his seminar for fruitful discussion of the present results. This work was supported in part by RFBR (Grants 05-02-16369 and 08-02-01376).

References

[1] Fulde, P.; Ferrell, R.A. *Phys. Rev.* 1964, A135, 550-563.
[2] Larkin, A.I.; Ovchinnikov, Yu.N. *Sov. Phys. JETP* Yu.N. 20, 762-771.
[3] Izyumov, Yu.A.; Proshin, Yu.N.; Khusainov, M.G. *Physics-Uspekhi* 2002, 45, 109-148.
[4] Buzdin, A.I. *Rev. Mod. Phys.* 2005, 77, 935-976.
[5] Goff, J.P. *et al. J. Magn. Magn. Mater.* 2002, 240, 592-594.
[6] Demler, E.A.; Arnold, G.B.; Beasley, M.R. *Phys. Rev. B* 1997, 55, 15174-15182.
[7] Abrikosov, A.A.; Gor'kov, L.P.; Dzyaloshinskii, I.E. *Quantum Field Theoretical Methods in Statistical Physics*, Pergamon Press, Oxford, 1965.
[8] Proshin, Yu.N.; Khusainov, M.G. *JETP* 1998, 86, 930-942.
[9] Eilenberger, G. *Z. Phys.* 1968, 214, 195-213.
[10] Abrikosov, A.A. *Fundamentals of the Theory of Metals*, North Holland, Amsterdam, 1988.
[11] Gor'kov, L.P.; Rusinov, A.I. *Sov. Phys. JETP* 1964, 19, 922-929.
[12] Hornreich, R.M.; Luban, M.; Strikman, S. *Phys. Rev. Lett.* 1975, 35, 1678-1681.
[13] Khusainov, M.G.; Proshin, Yu.N. *Physics-Uspekhi* 2003, 46, 1311-1313.
[14] de Gennes, P.G. *Rev. Mod. Phys.* 1964, 36, 225-237.

In: Progress in Superconductivity Research
Editor: Oliver A. Chang, pp. 103-121
ISBN: 1-60021-694-3

Chapter 4

THE SUPERCONDUCTIVITY IN CHEMICAL VAPOR DEPOSITION DIAMOND THICK FILMS

Changzhi Gu**, Zongli Wang, Dan Wu, Chunyan Li, Qiang Luo, Junjie Li, Nanlin Wang and Zengsun Jin

Beijing National Laboratory for Condensed Matter Physics, Institute of Physics, Chinese Academy of Sciences, Beijing 100080, China

Abstract

Heavily boron-doped polycrystalline diamond thick films were synthesized via the electron-assisted hot filament chemical vapor deposition method. The surface morphology and thickness for a freestanding polycrystalline diamond film were evaluated by scanning electron microscope (SEM), and the structure and phase purity of as-grown heavily boron-doped diamond film were characterized by X-ray diffraction (XRD) and micro-Raman spectroscopy. The film resistivity and the carrier concentration as well as the relation of the resistivity to carrier concentration were obtained at room temperature by the van der Pauw method and the Hall-effect method, respectively. Furthermore, the properties of electrical transport and magnetization were measured between room temperature and 1.9K, and the results indicated the superconducting transition temperatures were 10 K for onset and 8.3 K for zero resistance as well as the upper critical field was 2.7 T. Subsequently, we discussed the possible factors of inducing such a high Tc value for as-grown heavily boron-doped polycrystalline diamond film. In particular, the contrastive studies of optical response properties for present boron-doped diamond film and the undoped diamond film substantiated that there was the strong electron-phonon coupling in the heavily boron-doped diamond system.

1. Introduction

Diamond has many unique physical and chemical properties unmatchable by any other materials [1], such as wide band gap, negative electron affinity (NEA), high carrier mobility, extremely high breakdown field, highest hardness, high thermal conductivity, high mechanical strength and good corrosion resistivity, so it is an ideal candidate for a wide

* E-mail address: czgu@aphy.iphy.ac.cn. Corresponding author. Tel.: +86-10-82648197; Fax: +86-10-82649228

variety of applications. For example, it can be used in the heat sinks, cutting tools, abrasive structure components and bearing based on its excellent mechanical, thermal and chemical properties; meanwhile, it can be used as the X-ray, infrared radiation (IR) and laser window based on its corresponding optical properties [2, 3]. On the other hand, there are also interesting perspectives for diamond application as electronic devices in hostile, high-speed, high-density, high-voltage, or high-temperature environments, electron emitters, radiation sensors and particle detectors [4-7], which depend mainly upon its suitable electronic, thermal and chemical properties. In the past two decades, much interest has been focused on its electronic characterizations, transport and thermal properties.

It is well-known that an undoped diamond is an insulator with the wide band gap of 5.45 eV [8, 9]. When charge carriers are incorporated into a diamond, the resulting diamond will change from an insulator to a semiconductor and to a metal and eventually to a superconductor at higher doping concentration. With the purpose of finding out how impurities play a role in the superconductivity of diamond [10, 11], there are methods of adding dopants into pure diamond that have been reported: (1) doping during the growth process, (2) doping by ion implantation, and (3) doping by thermal diffusion [12]. In particular, the first method has been usually used to incorporate dopants into diamond, in which the chemical vapor deposition (CVD) method as well as high pressure high temperature (HPHT) synthesizing technique have been popularly used recently. On the other hand, it is well known to us that the boron atom has a smaller covalent radius than other potential dopants, and it is relatively readily introduced into the dense diamond lattice in high concentration. Therefore the boron is the most attractive impurity and acts as single electron acceptors [13, 14]. As the boron doping level increases, there is a gradual change from a semiconductor to a metal and eventually to a superconductor. It has been reported that the boron doping concentration in CVD synthetic diamond could arrive in 2×10^{21} cm^{-3} (1.1 at. %)(15, 16), 4.7×10^{21} cm^{-3} (2.7 at. %)(17), 8.5×10^{21} cm^{-3} (4.9 at. %)(18), and even more than 1.4×10^{22} cm^{-3} (7.1 at. %)(19). When the boron doping concentration is lower than 8×10^{20} cm^{-3} (0.5 at. %), the doped boron atoms usually substitute for the carbon atoms, and the substituted boron atoms are bonded to neighboring carbon atoms in the *sp*3 environment, and in the ground state the holes provided by boron atoms are bound in one of the threefold degenerate impurity states with a binding energy of 0.38 eV [16, 20]. With the boron doping concentration up to 7×10^{21} cm^{-3} (4 at. %), the doped boron atoms will occupy neutrally interstitial positions, and as the average distance between boron atoms is close to the acceptor Bohr radius, the metallic conduction will appear with the room temperature conductivity of 10^2 $(\Omega\ cm)^{-1}$ [21, 22].

Since the first experimental observation of superconductivity below about 4 K in heavily boron-doped polycrystalline diamond with the doping concentration of 4.9×10^{21} cm^{-3} (2.8 at. %) synthesized using a toroid high-press chamber at a pressure of 8-9 GPa and temperature of 2500-2800 K by Ekimov et al was reported in 2004 [23-25], this somewhat unexpected breakthrough has been encouraging attempts of many other researchers to prepare similar materials by means of more versatile plasma decomposition techniques. And very soon after, Bustarret et al from France reported the superconductivity in the (001) oriented single crystalline boron-doped diamond epilayers grown on the commercial (001)-oriented polished Ib substrates by the microwave plasma-assisted chemical vapor deposition (MPCVD) method [26, 27], in which the superconducting transition temperatures can be changed from 0 to 2.1

Table I. Reported experimental data on superconductivity in boron-doped diamonds. The asterisk (*) stands for the unpublished e-printed manuscript.

Sample type	n_B (cm^{-3})	Growth method	$T_{c1}(T_{c2})$ (K)	Reference	Year
Polycrystalline bulk samples	4.6×10^{21}	HPHT	4 (2.3)	23	2004
Polycrystalline bulk samples	2.5×10^{21}	HPHT	4.5 (2.2)	24	2004
Polycrystalline bulk samples	1.8×10^{21} - 2.5×10^{21}	HPHT	4.5 (3.4)	25	2004*
(001)-oriented single-crystalline thin films	3.6×10^{20} - 1.9×10^{21}	MPCVD	<0.05 -2.1 (1.9)	26, 27	2004
(111) oriented poly-crystalline thin films	9.4×10^{20}	MPCVD	7.4 (4.2)	28	2004
(111) oriented poly-crystalline thin films	4.2×10^{20} - 3.7×10^{21}	MPCVD	5.0 (1.7) - 8.7 (5.0)	29	2005
(111)-oriented single-crystalline thin films	8.4×10^{21}	MPCVD	11.4 (7.4)	18	2005*
(001)-oriented single-crystalline thin films	8.5×10^{21}	MPCVD	6.3 (3.2)	18	2005*
Polycrystalline diamond thick films	5.6×10^{20} - 1.2×10^{21}	MPCVD	4.5 (2.2)	30	2005
Polycrystalline bulk samples	4.6×10^{21}	HPHT	2.4 (1.4)	31	2006
Nanocrystalline diamond thin films	1.6×10^{21}	MPCVD	2.7 (1.7)	32, 33	2006
Highly (111) oriented polycrystalline thick films	7.3×10^{20}	HFCVD	10 (8.3)	our study	

K with boron doping contents varying from 2×10^{20} to 2×10^{21} cm^{-3} and the critical concentration for the onset of superconductivity was about ~ 5×10^{20} cm^{-3} which was close to the critical boron concentration for the metal-non-metal transition in this system. Meanwhile, Takano and Umezawa et al from Japan also successfully obtained superconductivity in the highly (111) oriented polycrystalline diamond thin films with boron contents from 3.0×10^{20} to 3.7×10^{21} cm^{-3} deposited using the MPCVD method on the silicon (001) substrates [28, 29], the superconducting transition temperatures in the sample of 3.7×10^{21} cm^{-3} boron contents are determined to be 8.7 K for on set and 5.0 K for zero resistance by transport measurement, and the corresponding upper crystal field is estimated to be around 7 T, which is significantly higher than that reported for HPHT diamond [23-25]. Besides, they also observed the superconducting transition temperatures of 11.4 K for on set and 7.4 K for off set in the homoepitaxially (111) oriented diamond films with a high boron doping concentration of 8.4×10^{21} cm^{-3} on the HPHT synthetic type Ib (111) diamond substrates [18]. Subsequently, Winzer et al from Germany also investigated in detail the electronic properties (including the resistivity, the magnetoresistivity, the Hall effect and the upper critical field) on the border between the normal and the superconducting state for the heavily boron-doped freestanding diamond films with 0.1-0.2 mm thickness prepared by MPCVD [30]. Recently, there are also some reports about the superconducting properties of polycrystalline boron-doped diamond synthesized at 2700K and higher pressure of 20 GPa using a multianvil apparatus [31] and boron-doped nanocrystalline diamond thin films with thicknesses of 60-500 nm using a MPCVD system [32,33]. In one word, the existence of superconductivity in heavily boron-doped diamond has been a well established effect, and the superconductivity of boron-doped diamond at several K has been the most interesting phenomenon [18, 23-33]. Here we summarize the reported experimental data on superconductivity (T_c) in heavily boron-doped diamond, as shown in Table I.

All above investigations from experimental studies have shown that boron-doped diamond is a typical type-II superconductor and the primary factor responsible for superconductivity can be the strong electron-phonon coupling in the heavily boron-doped diamond system. For this point, it has been also substantiated by the experimental results from the band dispersions of near the Fermi level obtained by performing the angle-resolved photoemission spectroscopy (ARPES) for the heavily boron-doped single-crystal diamond (111) film [34]: the overall occupied electronic structure directly relates to the metallic properties of heavily boron-doped superconducting diamond. On the other hand, there are lots of theoretical studies about superconductivity in boron-doped diamond using the method of first-principles calculations [35-42]. The foremost numerical and analytical calculations, which was based on the density functional [local-density approximation (LDA)] theory [35, 36] or the supercell approach [37, 38], revealed that boron-doped diamond is a three-dimensional MgB_2-type superconductor and its superconductivity is caused by the strong electron-phonon coupling that is the coupling of a few holes doped at the top of the bonding σ valence band to optical bond stretching phonon modes. The increase from two to three dimensions both reduces the density of states at the Fermi level and limits the strong softening of the optical modes mainly responsible for the high Tc in MgB_2 [39]. Therefore, Even if diamond had the same bare coupling constant as *MgB2*, which could be achieved with 10% doping, *Tc* would be only 25 K. Subsequently, some other reports about superconducting diamond detailedly investigated the electronic structure, lattice dynamics, and electron-phonon coupling of boron-doped diamond using various theoretical models, such as the Anderson s-d hybrid model [40], the pseudopotential plane-wave method [41], the full potential linear muffin-tin Orbital (FP-LMTO) approach based on the virtual crystal approximation (VCA) [42] and the local-spin-density-functional technique [43] so on. They also found a strong electron-phonon coupling in very agreement with the observed superconductivity in heavily boron-doped diamond. Such an electronic mechanism of band structure has been well supported by both the experimental observations [34-38] and theoretical studies [40-45]. These understands about the electronic structure of heavily boron-doped diamond may also be important for developing diamond-based devices that make use of the unique properties of diamond [34].

In this work, we prepared the heavily boron-doped polycrystalline diamond thick films using hot filament chemical vapor deposition (HFCVD) method. The morphology and structure of as-grown boron doped diamond thick films were characterized by scanning electron microscope (SEM), X-ray diffraction (XRD) and micro-Raman spectroscopy, and the film resistivity and the carrier concentration at room temperature are also measured by a Hall system. We observed the superconductive transition temperature of about Tc≈10 K by measuring the transport properties and alternating current (AC) magnetic susceptibility of heavily boron-doped polycrystalline diamond thick films. We also discussed the possible factors of inducing such a high Tc value in as-grown boron doped diamond thick films. On the other hand, the optical properties of polished boron-doped polycrystalline diamond film and undoped diamond film were also contrastively studied by measuring the near-normal incident reflectance spectra with a Bruker 66 *v*/ *s* spectrometer in the range from 40 to 50000 cm^{-1}, in order to provide the experimental evidence of strong electron-phonon coupling in heavily boron-doped diamond thick film.

2. Experimental Details

The heavily boron-doped polycrystalline diamond thick films were grown on a molybdenum substrate by using an electron-assisted hot filament chemical vapor deposition (EA-HFCVD) apparatus equipped a positive bias voltage system, as shown schematically in Fig. 1. The tantalum filament array of 6 Ta wires — with length of 100 mm, diameter of 0.5mm and spacing of 10mm — can be heated to 2000 - 2200 °C. The distance between the filament and substrate can be adjusted from 12-18 mm, the sample temperature was measured by a thermocouple mounted under the substrate. Before the deposition, the molybdenum substrate was firstly abraded by diamond powders with average grain sizes of 0.5 μm for 40 min in order to increase the diamond nucleation sites and improve diamond film growth on the substrates, and then cleaned with acetone and deionized water in an ultrasonic cleaner,

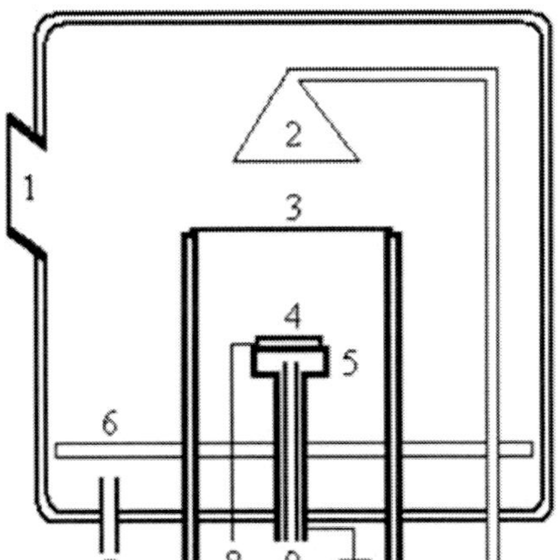

Figure 1. The schematic diagram of electron-assisted hot filament chemical vapor deposition system. (1) observation window, (2) Gas inlet: $CH_4 + H_2 + B(OCH_3)_3$, (3) Ta filament array, (4) Molybdenum substrate, (5) Cu holder, (6) adiabatic plate, (7) vacuum pump, (8) thermocouple, (9) cooling water, (10) filament electrical source, (11) bias voltage source.

respectively. The chamber of HFCVD apparatus was firstly filled with the mixture of methane (CH_4) and hydrogen (H_2), and the filaments are heated. When the temperature of filaments and substrate attain the preset values, a positive electrical potential is applied to the substrate via the substrate bias system, which can greatly enhance the growth rate of diamond film. The undoped diamond films begin to be grown and persist for about 1 hour, which is used to be a buffer layer of the boron-doped diamond film. This method can obtain the boron-doped diamond film of both good quality and high doping concentration. Subsequently, boron species were incorporated into the diamond films during the followed growth process by bubbling the H_2 gas through the Trimethyl borate ($B(OCH_3)_3$) liquid precursors at room temperature, and the boron-doped diamond films started being deposited. After growth duration of several ten hours, the sample was taken out and then the molybdenum substrates were removed by a cooling technique. Finally, a freestanding diamond thick film was obtained. The experimental parameters and its adjusting ranges were listed in Table П.

Table П. The deposition parameters of boron-doped diamond film in the EA-HFCVD system.

Experimental parameters	Adjusting range
Bias voltage	60-300 V
Bias current	1.5-15 A
Filament voltage	12 V
Filament current	240A
Number of filaments	3-7
Filament-substrate distance	12-18 mm
Filament temperature	2000-2200 °C
Substrate temperature	800-1000 °C
Total pressure	20-100 Torr
H_2 flow rate	100-500 sccm
CH_4 flow rate	2-10 sccm
$B(OCH_3)_3$ flow rate	0-30 sccm
Deposition time	40-150 h

For the as-grown boron-doped diamond film, the doping concentration and the quality were distinctly dependent on the deposition parameters, such as the substrate temperature, the total pressure, the bias voltage and the flow rates of H_2, CH_4 and $B(OCH_3)_3$. In particular, the flow rate ratios of H_2, CH_4 and $B(OCH_3)_3$ is the dominating factor in controlling the doping concentration of diamond film. For example, the diamond thick film with boron doping concentration of about 7.3×10^{20} cm^{-3} and thickness of 120 μm can be obtained under following parameters: a positive substrate bias voltage of 200 V; CH_4, H_2 and ($B(OCH_3)_3$) flow rates of 5 sccm, 200 sccm and 10 sccm; total pressure of about 50 Torr; filament and substrate temperatures of about 2000 °C and 850 °C; deposition duration of 80 hour. In above subscribed conditions, the growth rate of boron-doped diamond film is 1.5μm/h, which is less than that of undoped diamond film (6μm/h) under the same deposition condition.

The surface morphologies and thickness of as-grown heavily boron-doped diamond thick film were evaluated and measured with scanning electrical microscopy (SEM) of a focused ion beam (FIB) system. The film resistivity and the carrier concentration at room temperature were obtained in a Bio-Rad Microscience HL5500 Hall system. Especially, the film resistivity was estimated by the I-V measurements by the van der Pauw method, where the four probes were directly pressed on the surface of diamond film and formed a foursquare structure with area of 1 cm^2, and the carrier concentration was determined by the Hall-effect method, which is an average value from the boron-doped diamond film of 1 cm^2 and is calculated by the formulae:

$$U_H = KIB/d \ (K = 1/nq) \tag{1}$$

where U_H is the Hall voltage, d is the film thickness, n is the carrier concentration, q is the elementary charge, I and B stands for the DC current and the applied magnetic field, respectively. In addition, the structure and phase purity of the doped diamond samples were characterized by X-ray diffraction (XRD) with Cu K_α radiation and JY-6000 micro-Raman spectroscopy with the 532 nm laser wavelength, respectively.

In order to substantiate the superconductivity in a boron-doped diamond films, the transport properties of film were measured by a conventional four probe method using the PPMS 6000 bridge from American Quantum Design. 25-μm Pt wires were attached to the samples by silver epoxy. The measurements in the range of 1.9-300 K are performed in a 4He cryostat. AC magnetic susceptibility measurements were also performed by measuring the mutual inductance of a miniature coil with the PPMS 6000 bridge. For substantiating the strong coupling of electron with phonon being in heavily boron-doped diamond, on the other hand, the optical properties of polished boron-doped polycrystalline diamond film and undoped diamond film were also studied by measuring the near-normal incident reflectance spectra with a Bruker 66 *v/ s* spectrometer in the range from 40 to 25000 cm^{-1} and by a homemade grating spectrometer from 25000 cm^{-1} to 50000 cm^{-1}. The optical conductivity spectra were obtained by performing the Kramers-Kronig transformation of $R(\omega)$.

3. Results and Discussions

Fig. 2(a) and 2(b) show the SEM images of heavily boron-doped polycrystalline diamond thick film at the low magnification and high magnification, respectively. Compared with the undoped polycrystalline diamond thick film sample shown in Fig. 2(d), there are more obvious growth steps as well as clearer edges and corners for every crystallite in boron-doped diamond film. The average size of diamond grain is also much less than that of the undoped sample, although it is larger than 10 μm for the doped diamond thick film. The thickness of freestanding diamond film is estimated to be about 120 μm, which is obtained by observing its cross section SEM image shown in Fig. 2(c). On the other hand, the film resistivities and the carrier concentrations of as-grown boron-doped diamond thick films have been obtained by I-V measurements and Hall-effect measurements at room temperature, respectively. The

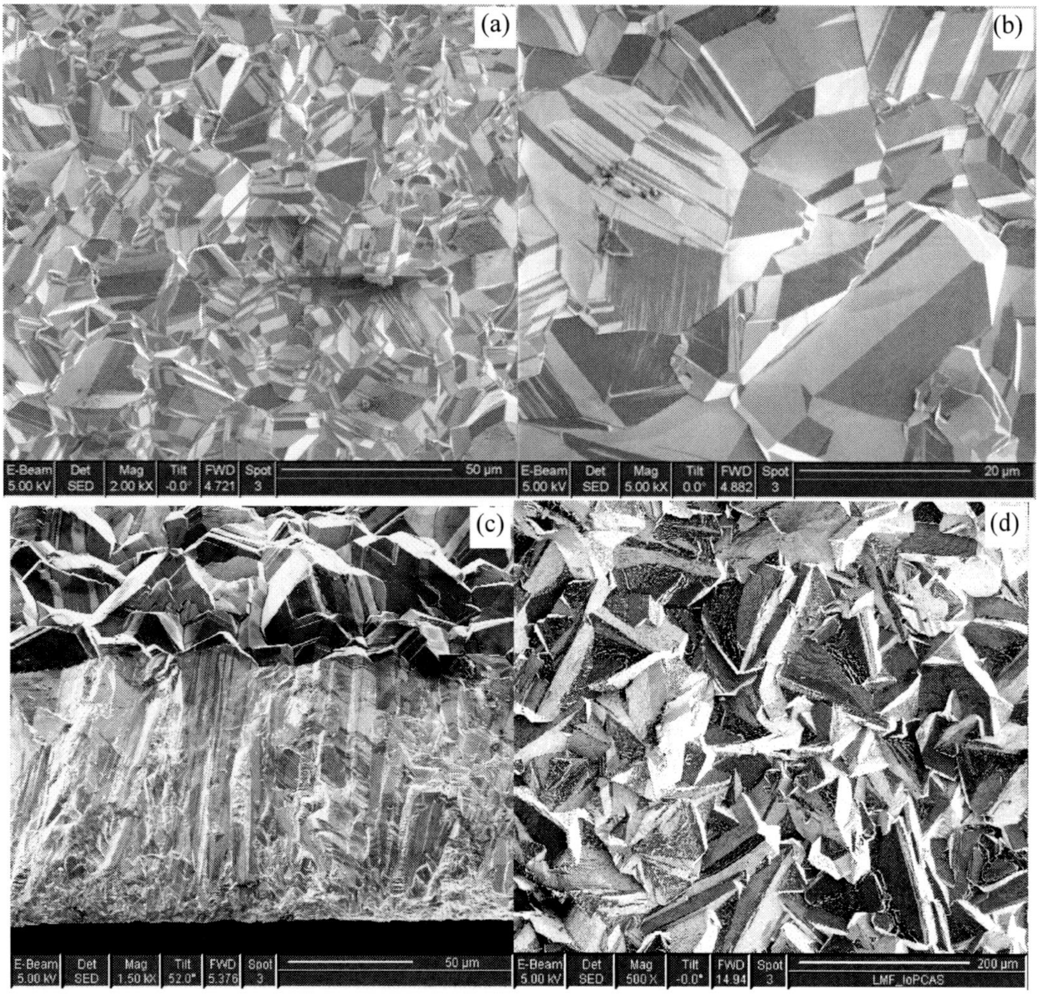

Figure 2. (a) Low-magnification and (b) high-magnification SEM images and (c) the cross section SEM images of the heavily boron-doped polycrystalline diamond thick film. (d) SEM image of undoped polycrystalline diamond thick film.

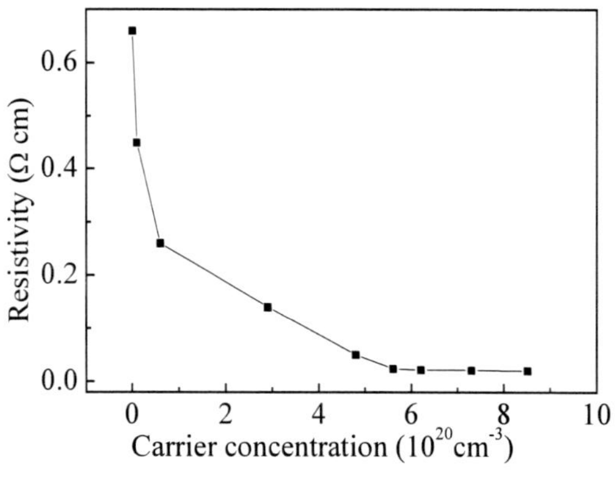

Figure 3. The plot of resistivity as a function of carrier concentration for the boron-doped diamond thick film samples.

plot of resistivity dependence of carrier concentration is shown in Fig. 3. It indicates that with the carrier concentration increasing from 7×10^{17} to 6×10^{20} cm^{-3}, the resistivity rapidly decreases from 0.65 to 0.024 Ω · cm; and the resistivity is basically changeless when the carrier concentration is more than 6×10^{20} cm^{-3} for the as-grown diamond samples. In the following, we will emphatically characterize the heavily doped diamond sample with the carrier concentration of 7.3×10^{20} cm^{-3}. In particular, the carrier concentration is estimated according formula (1) in the conditions of 19.9 mA DC current and 0.320 Tesla magnetic fields, and the measured film resistivity is about 0.0214 Ω · cm.

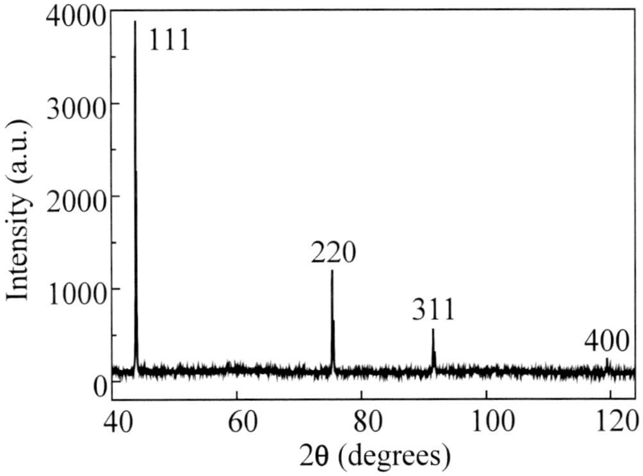

Figure 4. XRD diffraction spectrum of heavily boron-doped polycrystalline diamond thick film.

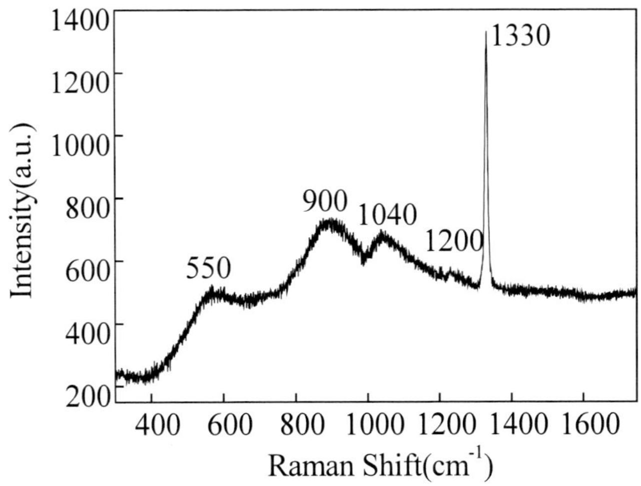

Figure 5. Micro-Raman spectrum of heavily boron-doped polycrystalline diamond film.

Crystal growth orientation is observed by means of the XRD for the polycrystalline diamond thick film with the doping concentration of 7.3×10^{20} cm^{-3}, as shown in Fig. 4. There are several peaks observed in the measured XRD spectrum. Firstly, the strongest sharp peak is detected at 2θ = 43.91°, corresponding to the (111) refraction of cubic cell of diamond structure. Secondly, there are three other detections at 2θ = 75.27°, 91.46° and 119.46°, which are the characteristics of the planes (220), (311) and (400), respectively. The intensities of these three peaks are obvious weaker than that of the (111) peak, moreover, the full-width half-maximum (HWHM) of (111) peak is only about 0.7°, which is the smallest in all the detections. Therefore, it suggests that the present doped diamond film has the fine crystalline structure with predominantly of (111) facts, which is consistent with the result from SEM observation shown in Fig.2. On the other hand, the Raman shifts of this heavily boron-doped polycrystalline diamond thick film is shown in the Fig. 5. Firstly, a negligible weak broad signal around 1500 cm^{-1} are detected in the Raman scattering spectrum, which indicates that there is very little content of amorphous carbon phase. Secondly, there is a sharp Raman peak at 1330 cm^{-1} resulting from the characteristic T_{2g} vibration mode of diamond. The zone-centre optical phonon line of diamond is normally observed at 1332 cm^{-1}, which is downshifted to be about 2 cm^{-1} and widened in the HWHM for the as-grown heavily doped diamond film. The downshift to lower wave numbers should be attributed to the uncertainty of the wave vector induced by the density of the boron pairs increasing in diamonds [46, 47]. Moreover, we can also find the asymmetry of the zone-center optical phonon line. The asymmetry should be due to the Fano interference induced by quantum mechanical interference between the discrete phonon state and the electronic continuum that results from the heavy boron doping. The phenomenon has been observed in the Raman spectra of heavily boron-doped CVD diamond films [48]. Finally, the Raman spectrum shows several large wide bands at approximately 570, 900, 1040 and 1200 cm^{-1} in the low frequency ranges, which have been typically observed in heavily boron-doped diamonds [47, 49-51]. The wide bands at both 900 and 1040 cm^{-1} and the weak broad band round 1200 cm^{-1} should be the Raman scattering spectra of (001) and (111) growth sectors, respectively [49]. In particular, these bands between 900 and 1200 cm^{-1} are attributed to the maxima in one phonon density of states, and they may come from the disordered regions, the phonon scattering at boron-induced structure modifications, or from boron-related electronic transitions and Raman scattering by phonons away from the center of Brillouin zone produced by heavy boron doping [49, 50]. For the broad Raman peak round 550 cm^{-1}, it may be associated to the local vibration modes of boron dimers [47]. These boron dimers have been reported to be related with some of the gap states, and that they are both electrically and vibrationally inactive with respect to the electron-phonon coupling driving the superconducting transition in metallic diamond [51]. As a result, we can conclude that the presence of these wide signals at the low frequency maybe correlate with the superconductivity of heavily boron-doped diamonds.

The transport properties of heavily boron-doped polycrystalline diamond film have been measured from room temperature (RT) to 1.9K, and the result was compared with that of undoped diamond film. Fig. 6(a) shows the electrical resistivity characteristics of the heavily boron-doped and the undoped polycrystalline diamond thick films under the zero magnetic field condition as a function of temperature, corresponding to the black square-lines and the red circle-lines, respectively. Firstly, the resistivty of doped diamond sample is of the order of 10^{-2} $\Omega \cdot$ cm, which is much lower than surface resistivity of undoped diamond sample (~10^{5} $\Omega \cdot$ cm). Secondly, the resistivities of both doped and undoped diamond films are gradually increases with the decrease of temperature. For the doped diamond sample, in particular, its resistivity begins to decrease rapidly near 10 K until it drops to zero within an accuracy of 10 μΩ at around 8.3 K, corresponding to the onset (T_c onset) and offset (T_c offset) of a superconducting transition, respectively. This substantiates the superconductivity in the heavily boron-doped diamonds, as reported by many other publications [23-33]. Fig. 6(b) also shows the magnetic susceptibility of heavily boron-doped diamond film near the superconducting transition temperature, which was measured at a fixed magnetic field 0.2 Oe. It displays that the magnetization′ begins to change near 8.9 K, indicating a diamagnetic response in the magnetic susceptibility, which also identified a superconducting transition in the present doped diamond sample. It is obvious that there are a few differences between the transition temperature values determined by the measurements of resistivity and magnetic susceptibility, and the former (10K) is slightly higher than the latter (8.9K), which is probably due to the relatively large amplitude of the oscillating field [52].

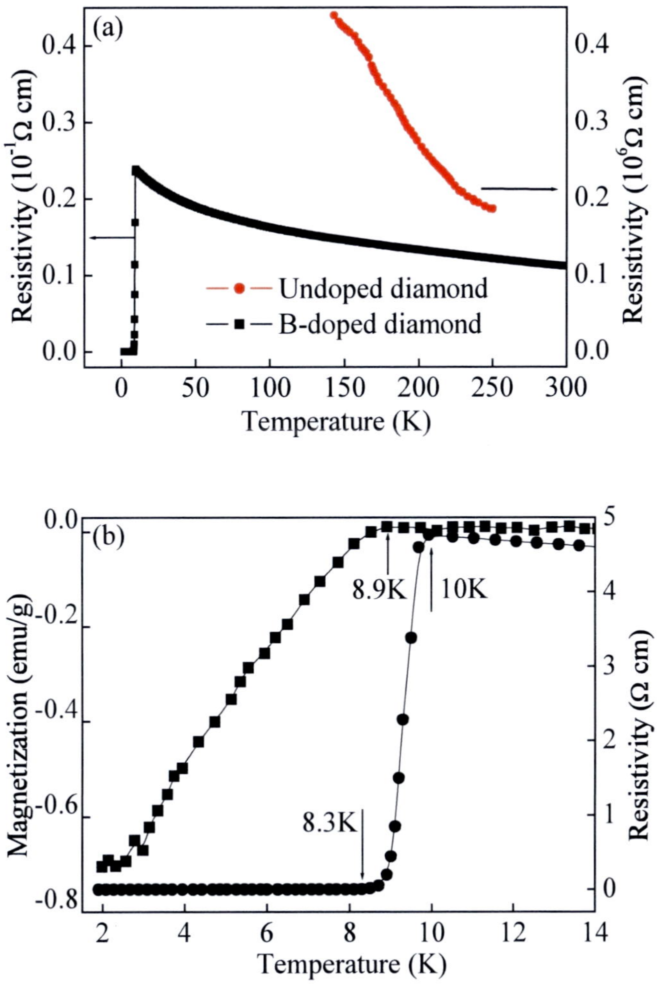

Figure 6. (a) The temperature dependence of the electrical resistivity of boron-doped and undoped polycrystalline diamond thick films at the zero magnetic fields. (b) The temperature dependence of the electrical resistivity and AC magnetic susceptibility of boron-doped diamond thick film near the superconducting transition temperature.

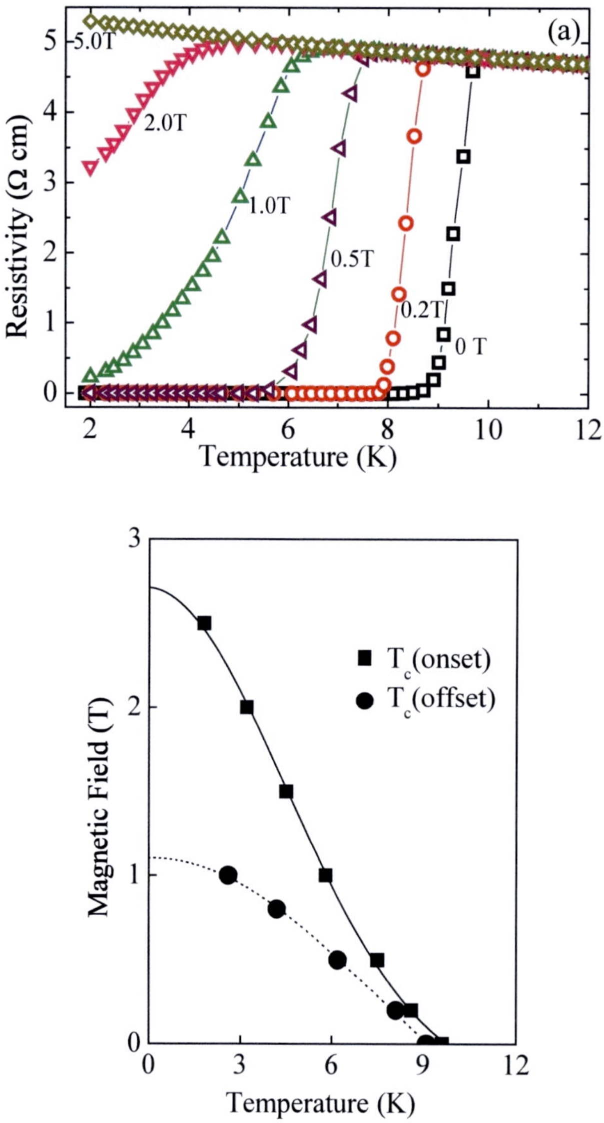

Figure 7. (a) The temperature dependence of film resistivity between 2 and 12 K under gradually increasing external magnetic field up to 5 T. (b) The applied magnetic field dependence of superconducting transition temperature. The solid squares (circles) have being deduced from temperature sweeps of the electrical resistivity for $R / R_N = 90\%$ ($R / R_N = 10\%$).

For investigating the external magnetic field effect on superconducting transition for the boron-doped diamond superconductor, the plots of the temperature dependence of film resistivity near T_c values under gradually increasing external magnetic field up to 5 T have been measured, as shown in Fig. 7(a). It shows clearly that with the magnetic field increasing, the values of T_c onset and T_c offset both shift to lower values, and the transition width (corresponding to the difference between the T_c onset and T_c offset) remains relatively small up to about 0.5T and rapidly increases for larger magnetic field. In order to define an accurate H_{c2} (T) line, the classical criterion has been cited, which is defined to be $R / R_N = 90\%$, where

R_N being the normal classical resistance [26]. Based on the classical criterion, the superconducting transition temperatures as a function of magnetic field have been drawn in Fig. 7(b). It indicates that the upper critical field $H_{c2}(0)$ is about 2.7 T, which is slight larger than the estimated value of 2.1 T calculated with the standard relationship for a dirty type-II superconductor [53]: $H_{c2}(0) = -0.69T_c\left[dH_{c2}(T)/dT\right]\Big|_{T_c}$. In particular, the $H_{c2}(0)$ value is related to the Ginzburg-Landau coherence length (ξ) according to the relation [23]: $\xi = [\Phi_0/2\pi H_{c2}(0)]^{1/2}$. Moreover, it's well known to us that the $H_{c2}(0)$ value is smaller than the reported by other references [23-25, 28, 29], but is close to or lager than the reported value by Refs. 26, 30 and 31, which suggests that the ξ value may be different in the different diamond samples and hence results in the fluctuation of upper critical field. On the other hand, the line corresponding to the criterion of $R / R_N = 10\%$ are also shown in Fig. 7(b), It indicates the increase of transition width when the applied magnetic field is larger than 0.5 T, which means the presence of quantum fluctuations in the present doped diamond system [26].

Here we summarized our results and other references about the superconductivity of boron-doped diamonds in Table I. It is clear that the superconducting transition widths are found to be larger than 1K and even up to 4K for almost polycrystalline or single crystalline diamond samples, which is an indication of chemical inhomogeneity of the samples. And besides, some other problems also appear in the CVD diamond samples due to the effects of complex surface. For example, the reported value of Tc onset ranges from <0.05 to 11.4 K. However, there is not any clear correlation between the boron concentration and the superconducting transition temperature in the published data. Until to now, the only reproducible observation is the fact of superconductivity in boron-doped diamonds. Our experimental results have also shown that the heavily boron-doped polycrystalline diamond is a typical type-II superconductor, and the T_c onset and T_c offset is estimated to be about 10 K and 8.3 K, respectively. For the fact of the higher Tc value in as-grown polycrystalline diamond thick film with the relative low doping concentration of 7.3×10^{20} cm^{-3}, There are several factors to be considered. Firstly, it has been substantiated by Minhaj *et al* that comparing with thin polycrystalline film material, thicker polycrystalline film or bulk materials can induce a higher superconducting transition temperature due to the proximity effect and weak localization [54]. The present film is estimated to be the thickness of about 120 μm, and can be considered as a bulk material and be of higher superconducting transition temperature. However, it may be only a secondary factor. The most important factor induced the larger T_c value should be the very strong electron-phonon coupling in the present boron-doped diamond sample. Of course, there have been many reported reasons for strong electron-phonon coupling in the heavily doped diamonds, such as the higher efficiency of doping in samples [28], the existence of moderate Coulomb repulsion and disorder in bulk [55], the contraction of the reconstructed bonds and the 2D nature of the surface states [35-37] for the boron-doped diamond samples. These theoretical assumptions might be the factor in the strong electron-phonon coupling. However, the directly confirmation of strong electron-phonon coupling should be made by the experimental observations. Here the coupling of electron with phonon has been experimentally proved by our recent infrared response study of present boron-doped diamond film, which will be followed by the discussion about the optical properties of boron-doped diamond sample.

The SEM images shown in Fig. 2 have indicated that there are different growth orientations for diamond grains in the heavily boron-doped polycrystalline diamond thick film, and hence the polycrystalline diamond films have very rough surfaces, which isn't suit for the optical reflectance measurement. As the optical experiment needed, the sample should be a smooth and shiny surface as soon as possible. Therefore the surface of as-grown diamond thick film has to be polished, where a high rotate-speed (20000 rounds / min) diamond grinding wheel has been used to polish the sample and then a mirror polished surface has been obtained. Meanwhile, the undoped diamond thick film has also been polished in order to form a mirror polished surface. Their optical properties have been comparatively studied by measuring the near-normal incident reflectance spectra with a Bruker 66 *v/ s* spectrometer in a wide frequency range of 40–50000 cm^{-1}, where an *in situ* overcoating technique was employed for reflectance measurements [56]. Fig. 8 gives a schematic picture of as-employed classic reflex spectral apparatus, where the path of the incident light is drawn and the sample is mounted on an optically black cone in a cold-finger flow cryostat. The light beam starts off from the light source, after passing through the Michelson's interferometer, it will convergently irradiate to the crystal sample surface with an incident angle of about 8 degrees. The incident light beam will be reflected by the sample surface; finally, it comes into the detectors of D1, D2, or D3, where D1, D2, and D3 are the detectors with different detecting wave ranges of near infrared and visible lights, middle infrared light and far infrared light, respectively.

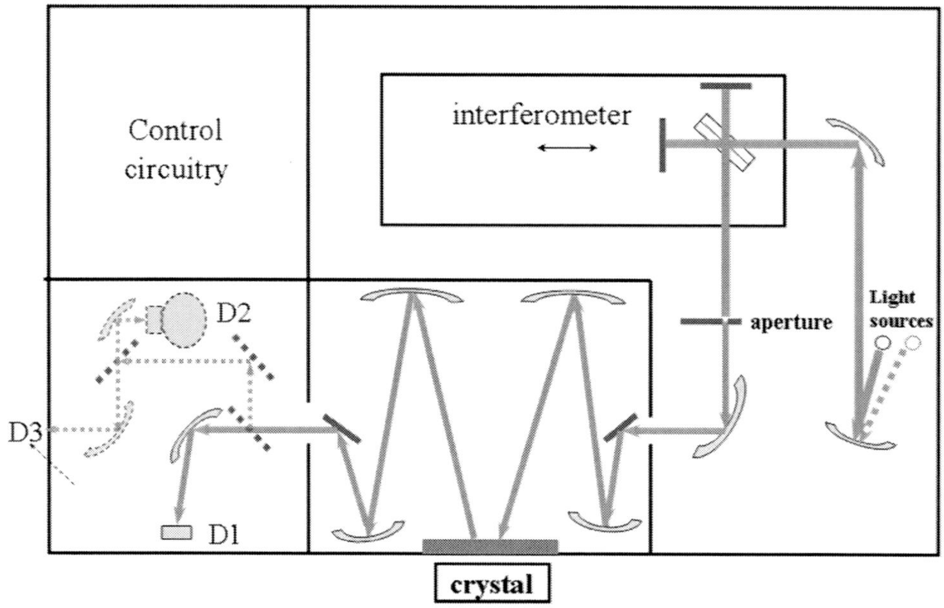

Figure 8. The schematic diagram of measurement configuration employed for the near-normal incident reflectance spectra measurement configuration, where the path of the incident light is drawn and the sample is mounted on an optically black cone in a cold-finger flow cryostat.

The measured optical reflectance spectra of doped and undoped diamond samples from 40 to 8500 cm^{-1} at RT and 10 K were shown in the Fig. 9. As for the undoped diamond sample, the reflectance spectrum shows featureless change and is basically independent of temperature over a wide frequency range. On the other hand, there is weak temperature effect on the reflectance spectrum for the boron-doped diamond sample. Firstly, the reflectivity is independent of temperature when the frequency is above 6400 cm^{-1}. Secondly, when the frequency of incident light is between 2400 and 6400 cm^{-1} or is smaller than 150 cm^{-1}, the measured reflectivity decreases with the decrease of temperature, it indicates an insulating characteristic. However, it gradually increases with the temperature decreasing for the reflectivity in the frequency range from 150 to 2400 cm^{-1}. Furthermore, it clear from Fig. 8 that there is a major difference between the boron-doped and undoped diamond samples at the frequency below 5000 cm^{-1}, which is an abruptly increase of the reflectivity for the doped diamond sample. It illustrates clearly the formation of electronic states at the low energies due to boron-doping. In order to lucubrate the electronic states induced by the boron-doping, the Kramers-Kronig transformation of $R(\omega)$ have been performed to obtained the optical conductivity spectra, where the constant and Hagen-Rubens relation is used for the doped and undoped diamond samples, respectively. Due to the very weak dependent of the conductivity spectra on the temperature, we only gives the optical conductivity spectra of doped and undoped diamond samples at the room temperature, as shown in the inset of Fig. 9. Compared with the undoped diamond sample, the weight of conductivity spectra for boron doped diamond sample is an evident increase at frequency below 7000 cm^{-1}. In particular, the low energy excitations do not form a Drude peak. Interesting, the conductivity was severely suppressed at low frequency range, and resulting a broad peak appears at about 3000 cm^{-1}, its peak energy is identical with the boron acceptor levels located at 0.38 eV from the top of valence band, as determined from many other experimental studies [20, 57], which suggests that the states induced by boron-doping are still highly localized. And hence the peak energy exactly corresponds to the interband transition from the top of the valence band to the impurity states.

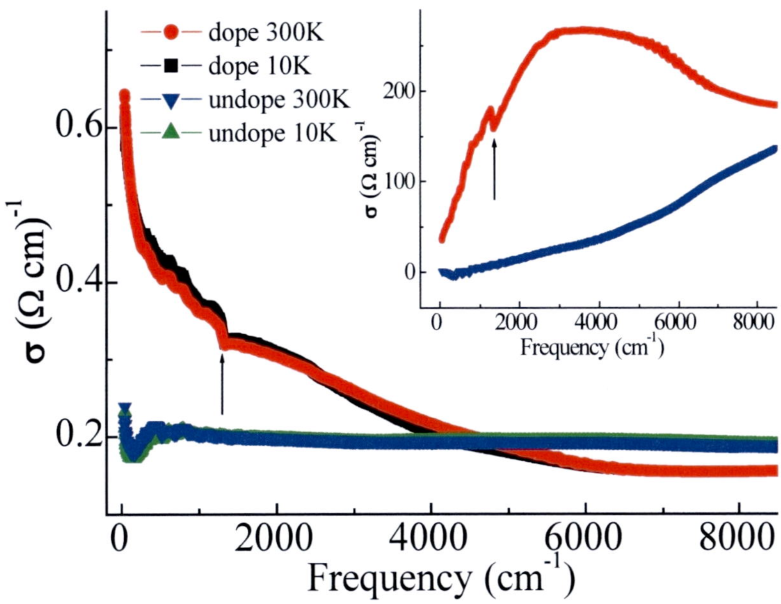

Figure 9. The frequency dependence of optical reflectance spectra of boron-doped diamond and undoped diamond in the frequency range in the range of 40–8500 cm^{-1} at the temperatures of 300 K and 10 K, respectively. The inset shows the optical conductivity spectra of pure and boron-doped diamond at room temperature.

From the above experimental results, we can obtain the evolution information of the electronic states for the boron doped diamond with doping concentration. Firstly, the undoped diamond should be a standard semiconductor, of which the valence band is completely filled, the conduction band is completely empty, and the Fermi level locates close to the top of the valence band. Secondly, the acceptor impurity states start to be created near the Fermi level with a binding energy of 0.38 eV due to the boron-doping. Moreover, these completely localized impurity energy levels could be gradually broaden and then formed the band with the increase of boron-doping concentration. Finally, the impurity band can cross the Fermi level and the metallic conduction could be produced when the doping level is enough high. Of course, a slight downward shift of Fermi level likely occurs due to the holes created in the valence band, which can be used to explain the as-observed energy shift of interband transitions obtained by the optical conductivity spectra of boron-doped diamond. Furthermore, the superconductivity should originate from the boron-doping induced impurity band, corresponding to the strong electron-phonon coupling in the heavily boron-doped diamonds.

Boron-doping does not only result in the change of electronic states but also produce an important difference between the doped and undoped diamond samples. It is the peculiar structure at about 1330 cm^{-1} in the optical reflectivity and conductivity spectra of the boron-doped diamond sample as shown in Fig.9 (presented with arrow), but there is no anomaly of the spectra to be detected for the undoped sample. It's well known that there is a very sharp zone-center phonon peak centered at 1332 cm^{-1} in the Raman spectra of undoped diamonds [48], which are obviously forbidden in infrared region due to the featureless of its reflectance. For the boron-doped diamonds crystal structure, a part of carbon atoms are substituted by the boron atoms, the resulting distortion of vibration eigenvectors brings nonzero moments of electronic dipoles. Therefore, the infrared forbidden phonon mode is activated by boron-doping and detected at about 1330 cm^{-1} in the spectra of reflectivity and conductivity for the boron-doped diamond. In particular, this phonon mode has an anti-resonance characteristic, looking more like an asymmetric dip than a peak in the electronic background of conductivity spectrum. The rather obvious anti-resonance feature suggests the strong electron-phonon coupling in the boron-doped diamond system, which has been usually observed in the organic conducting or superconducting materials due to this kind of significant electron-molecular vibration coupling [58, 59]. The presence of such a peculiar phonon structure clearly indicates a very strong electron-phonon coupling in the as-grown boron-doped diamond sample. As a result, such infrared response study can provide an experimental evidence for the strong coupling of electrons with this mode. Moreover, the strong coupling and its significant effect on the superconductivity in boron-doped diamond have been recently mentioned in many theoretical studies [35-42].

4. Conclusion

We firstly present unequivocal substantiation for superconductivity in the boron-doped polycrystalline diamond thick film prepared by EA-HFCVD method. The results from the measurements of resistivity and magnetic susceptibility show that it is a typical type-II superconductor and the superconducting transition temperature is 10 K for T_c onset and 8.3 K for zero resistance. The fact of higher T_c value in as-grown polycrystalline diamond film can be primarily attributed to the existence of very strong electron-phonon coupling induced by a combination of the higher efficiency of doping, the contraction of the reconstructed bonds or the 2D nature of the surface states. On the other hand, the strong electron-phonon coupling is substantiated by the studies of optical reflective and conductive properties on polished surfaces of boron-doped diamond films, where it clearly illustrates the formation of an acceptor impurity state near the top of the valence band as well as the impurity's effect on phonon vibration modes and the quite strong electron-phonon coupling in such a heavily boron-doped diamond system.

Acknowledgments

This work was supported by the National Natural Science Foundation of China (Grant No.50672121), the State Key Development Program for Basic Research of China (Grant No.2002CB613500 and 2003CB716900), the National High technology Research and Development Program of China (Grant No.2002AA325090) and National Center for Nanoscience and Technology, China.

References

[1] Field, J. E. In *The Properties of Natural and Synthetic Diamond*, Field, J. E.; Ed.; Academic Press, New York, 1992; pp 3-34.
[2] Woerner, E.; Wild, C.; Sebert, W. M.; Koidl, P. *Diamond Relat. Mater.* 2001, 10, 557-560.
[3] Isberg, J.; Hammersberg, J.; Johansson, E.; Wikström, T.; Twitchen, D. J.; Whitehead, A. J.; Coe, S. E.; Scarsbrook, G. A. *Science* 2002, 297, 1670-1672.
[4] Vescan, A.; Daumiller, I.; Gluche, P.; Ebert, W.; Kohn, E. *Diamond Relat. Mater.* 1998, 7, 581-584.
[5] Manfredotti, C.; Giudice, A. L.; Ricciardi, C.; Paolini, C.; Massa, E.; Fizzotti, F.; Vittone, E. *Nucl. Instrum. Methods A* 2001, 458, 360-364.
[6] Bergonzo, P.; Tromson, D.; Mer, C.; Guizard, B.; Foulon, F.; Brambilla, A. *Phys. Status Solidi A* 2001, 185, 167-181.
[7] Reichart, P.; Datzmann, G.; Hauptner, A.; Hertenberger, R.; Wild, C.; Dollinger, G. *Science* 2004, 306, 1537-1540.
[8] Walker, W. C.; Osantowski, J. *Phys. Rev.* 1964, 134, A153-A157.
[9] Saslow, W.; Bergstresser, T. K.; Cohen, M. L. *Phys. Rev. Lett.* 1966, 16, 354-356.
[10] Shiomi, H.; Nishibayashi, Y.; Fujimori, N. *Jpn. J. Appl. Phys.* 1991, 30, 1363-1366.
[11] Lagrange, J. P.; Deneuville, A.; Gheeraert, E.; *Diamond Relat. Mater.* 1998, 7, 1390-1393.
[12] Thonke, K. *Semicond. Sci. Technol.* 2003, 18, S20-S26.
[13] Chrenko, R. M. *Phys. Rev. B* 1993, 7, 4560-4567.
[14] Kalish, R. *Diamond Relat. Mater.* 2001, 10, 1749-1755.
[15] Brunet, F.; Germi, P.; Pernet, M.; Deneuville, A.; Ghaeeraert, E.; Laugier, F.; Burdin, M.; Rolland, G. *Diamond Relat. Mater.* 1998, 7, 869-873.
[16] Bustarret, E.; Ghaeeraert, E.; Watanabe, K. *Phys. Status Solidi A* 2003, 199, 9-18.

[17] Hartmann, P.; Bohr, S.; Haubner, R.; Lux, B.; Wurzinger, P.; Griesser, M.; Bergmaier, A.; Dollinger, G.; Sternschulte, H.; Sauer, R. *Int. J. Refract. Met. Hard Mater.* 1997, 16, 223-232.

[18] Umezawa, H.; Takenouchi, T.; Takano, Y.; Kobayashi, K.; Nagao, M.; Sakaguchi, I.; Tachiki, M.; Hatano, T.; Zhong, G. F.; Tachiki, M.; Kawarada, H. *cond-mat* / 0503303 (unpulished).

[19] Sawamura, A.; Iihara, J.; Muramatsu, Y.; Takebe, T.; Namba, A.; Perera, R. C. C. *Abstracts of the Tenth International Conference on New Diamond Science and Technology*, Tsukuba, Japan, 9–12 May 2005, pp. 57.

[20] Nakamura, J.; Kabasawa, E.; Yamada, N.; Einaga, Y.; Perera, R. C. C. *Phys. Rev. B* 2004, 70, 245111-1-6.

[21] Werner, M.; Dorsch, O.; Baerwind, H. U.; Obermeier, E.; Haase, L.; Seifert, W.; Ringhandt, A.; Johnston, C.; Romani, S.; Bishop, H.; Chalker, P. R. *Appl. Phys. Lett.* 1994, 64, 31-33.

[22] Werner, M.; Dorsch, O.; Baerwind, H. U.; Obermeier, E.; Haase, L.; Seifert, W.; Ringhandt, A.; Johnston, C.; Romani, S.; Bishop, H.; Chalker, P. R. *Appl. Phys. Lett.* 1994, 64, 595-597.

[23] Ekimov, E. A.; Sidorov, V. A.; Bauer, E. D.; Mel'nik, N. N.; Curro, N. J.; Thompson, J. D.; Stishov, S. M. *Nature* 2004, 428, 542-545.

[24] Sidorov, V. A.; Ekimov, E. A.; Bauer, E. D.; Mel'nik, N. N.; Curro, N. J.; Fritsch, V.; Thompson, J. D.; Stishov, S. M.; Alexenko, A. E.; Spitsyn, B. V. *Diamond Relat. Mater.* 2005, 14, 335-339.

[25] Sidorov, V. A.; Ekimov, E. A.; Stishov, S. M.; Bauer, E.D.; Thompson, J. D. *cond-mat* / 0409624 (unpulished).

[26] Bustarret, E.; Kačmarčik, J.; Marcenat, C.; Gheeraert, E.; Cytermann, C.; Marcus, J.; Klein, T. *Phys. Rev. Lett.* 2004, 93, 237005-1-4.

[27] Kačmarčik, J.; Marcenat, C.; Cytermann, C.; Silva, A. F.; Ortega, L.; Gustafsson, F.; Marcus, J.; Klein, T.; Gheeraert, E.; Bustarret, E.; *Phys. Status Solidi, A* 2005, 202, 2160-2165.

[28] Takano, Y.; Nagao, M.; Sakaguchi, I.; Tachiki, M.; Hatano, T.; Kobayashi, K.; Umezawa, H.; Kawarada, H. *Appl. Phys. Lett.* 2004, 85, 2851-2853.

[29] Takano, Y.; Nagao, M.; Takenouchi, T.; Umezawa, H.; Sakaguchi, I.; Tachiki, M.; Kawarada, H. *Diamond Relat. Mater.* 2005, 14, 1936-1938.

[30] Winzer, K.; Bogdanov, D.; Wild, C. *Phys. C* 2005, 432, 65-70.

[31] Dubrovinskaia, N.; Eska, G.; Sheshin, G. A.; Braun, H. *J. Appl. Phys.* 2006, 99, 033903-1-7.

[32] Nesládek, M.; Tromson, D.; Mer, C.; Bergonzo, P.; Hubik, P.; Mares, J. J. *Appl. Phys. Lett.* 2006, 88, 232111-1-3.

[33] Mareš, J. J.; Nesládek, M.; Hubík, P.; Kindl, D.; Krištofik, J. *Diamond Relat. Mater.* 2007, 16, 1-5.

[34] Yokoya, T.; Nakamura, T.; Matsushita, T.; Muro, T.; Takano, Y.; Nagao, M.; Takenouchi, T.; Kawarada, H.; Oguchi, T. *Nature* 2005, 438, 647-650.

[35] Boeri, L.; Kortus, J.; Andersen, O. K. *Phys. Rev. Lett.* 2004, 93, 237002-1-4.

[36] Lee, K. W.; Pickett, W. E. *Phys. Rev. Lett.* 2004, 93, 237003-1-4.

[37] Blase, X.; Adessi, C.; Connétable, D. *Phys. Rev. Lett.* 2004, 93, 237004-1-4.

[38] Xiang, H. J.; Li, Z. Y.; Yang, J. L.; Hou, J. G.; Zhu, Q. S. *Phys. Rev. B* 2004, 70, 212504-1-4.
[39] Liu, A.Y.; Mazin, I. I.; Kortus, J. *Phys. Rev. Lett.* 2001, 87, 087005-1-4.
[40] Pogorelov, Y. G.; Loktev, V. M. *Phys. Rev. B* 2005, 72, 075213-1-4.
[41] Tse, J. S.; Ma, Y.; Tutuncu, H. M. *J. Phys.: Condens. Matter* 2005, 17, S911-S920.
[42] Goss, J. P.; Briddon, P. R. *Phys. Rev. B* 2006, 73, 085204-1-8.
[43] Boeri, L.; Kortus, J.; Andersen, O. K. *J. Phys. Chem. Solids* 2006, 67, 552-556
[44] Yanagisawa, H.; Tanaka, T.; Ishida, Y.; Rokuta, E.; Otani, S.; Oshima C. *Phys. Rev. B* 2006, 73, 045412-1-7.
[45] Sacépé, B.; Chapelier, C.; Marcenat, C.; Kačmarčik, J.; Klein, T.; Bernard, M.; Bustarret, E. *Phys Rev Lett.* 2006, 96, 097006-1-4.
[46] Zhang, R. J.; Lee, S. T.; Lam, Y. W. *Diamond Relat. Mater.* 1996, 5, 1288-1294.
[47] Bernard, M.; Baron, C.; Deneuville, A. *Diamond Relat. Mater.* 2004, 13, 896-899.
[48] Ager, J. W.; Walukiewicz, W.; McCluskey, M.; Plano, M. A.; Landstrass, M. I. *Appl. Phys. Lett.* 1995, 66, 616-618.
[49] Wang, Y. G.; Li, H. D.; Lin, Z. D.; Feng, K. *Jpn. J. Appl. Phys.* 2000, 39, 2795-2976.
[50] Gonon, P.; Gheeraert, E.; Deneuville, A.; Fontaine, F.; Abello, L.; Lucazeau, G. *J. Appl. Phys.* 1995, 78, 7059-7062.
[51] Bourgeois, E.; Bustarret, E.; Achatz, P.; Omnès, F.; Blase, X. *Phys. Rev. B* 2006, 74, 094509-1-8.
[52] Watson, J. H. P. *Phys. Rev.* 1966, 148, 223-230.
[53] Werthamer, N. R.; Helfand, E.; Hohenberg, P. C. *Phys. Rev.* 1966, 147, 295-302.
[54] Minhaj, M. S. M.; Meepagala, S.; Chen, J. T.; Wenger, L. E. *Phys. Rev. B.* 1994, 49, 15235-15240.
[55] Baskaran, G. *cond-mat* / 0404286 (unpulished).
[56] Homes, C. C.; Reedyk, M.; Crandles, D. A.; Timusk, T. *Appl. Opt.* 1993, 32, 2976-2983.
[57] Barnard, A. S.; Russo, S. P.; Snook, I. K. *Philos. Mag.* 2003, 83, 1163-1174.
[58] Jacobsen, C.S. In *Semiconductors and Semimetals*; Conwell, E.; Ed.; Academic Press, London, 1998; Vol. 27, pp. 293-298.
[59] Wang, N. L.; Mori, H.; Tanaka, S.; Dong, J.; Clayman, B. P. *J. Phys.: Condens. Matter* 2001, 13, 5463-5470.

In: Progress in Superconductivity Research
Editor: Oliver A. Chang, pp. 123-162 ISBN 1-60021-694-3

Chapter 5

DDW ORDER AND ITS ROLE IN CUPRATES

Angsula Ghosh
Manaus, Brazil

Abstract

The high-T_c cuprates have been a focus of intense study since its discovery in 1986. The 20th anniversary of the discovery of this material is marked by a partial understanding of its properties. They are far from being fully understood. The cuprates not only demonstrate a very high T_c, but, also exhibit various anamolies in their superconducting and the normal state properties. The pairing symmetry of the Cooper pairs was a topic of intense debate for more than a decade. A new class of phase sensitive tests as well as various other techniques for energy-gap measurements have revealed a predominantly d-wave pairing symmetry in the cuprates unlike the s-wave states in conventional superconductors. At present one of the most peculiar and controversial topics in these materials is the phenomenon of pseudogap in the normal state of the cuprates. A depletion of the low-energy density of states is observed in the underdoped regime of these materials by various experimental measurements like the optical conductivity, ARPES, c-axis tunneling, specific heat etc. The pseudogap has also been observed to have the same k-dependence as that of the superconducting gap. A number of theoretical scenarios have been proposed to explain the origin of this phase. It was initially believed to be a crossover phenomenon where the pseudogap had been ascribed as a natural consequence of local, dynamic pairing correlations in the normal state of low-density small pair size superconductors. Apart from the other suggestions like the phase fluctuations of the superconducting order parameter in the pseudogap regime, recently, the d-densiy wave (DDW) scenario has been proposed as a strong candidate to explain the pseudogap. In fact, various theoretical works on DDW have been successful in explaining some of the salient features of this phase. A number of experiments are also consistent with this proposal. In this article we will present a brief overview of the phase diagram of the cuprates and the present experimental situation followed by a discussion on the present status of the DDW order as a candidate for the pseudogap. We will present a comparison between the DDW order and the other scenarios. We will also explain,in short, the conventional density-wave order and its difference with the unconventional or DDW order. We will discuss the theoretical findings of the DDW order in the limelight of the available experimental predictions. Some proposals of experimental measurements which might play an important role to strengthen the candidature of DDW order will also be considered.

1. Introduction

In recent decades the study of the high temperature superconductors (HTSC) has been one of the central issues both in experimental and theoretical research fields of condensed matter physics. The discovery of superconductivity in cuprates in 1986 [1], which eventually gave birth to the area of high temperature superconductivity, and the subsequent rapid rise in the critical temperatures above the melting point of nitrogen [2] have led to a phenomenal rise in the literature. For years prior to the discovery of HTSC, the highest T_c was at 23K. With the appearence of the cuprates, not only a quantum leap was observed in the critical temperatures but also the high T_c's were observed in totally different kinds of compounds, namely the transition metal oxides. A number of experimental results which are not only anamolous but also lack an accordance among themselves made it difficult to construct a microscopic theory of HTSC. However, it is clear that some different mechanism was is work. Hence, the theory of the high temperature superconductors is far from being resolved. The high critical temperatures (well above the melting point of nitrogen) ushered in an era of great excitement due to the great strides both in the technological application and basic research.

The high temperature superconductors are different from the conventional BCS (Bardeen, Cooper and Schrieffer) [3] superconductors due to a wide range of its properties, both in the normal (metallic state) and superconducting state of the material. The discovery of superconductivity in the transition metal oxides made it clear that the physics of these materials are very distinct from that of the conventional ones. These new materials are all layered compounds with one or more copper-oxygen planes and are more commonly called cuprates. It is widely believed that the source of the pairing interaction which is responsible for the superconducting transition lies in the CuO_2 planes and hence could be explained by two-dimensional models. They all satisfy a very rich phase diagram which we discuss in short before going into further details. The parent compound is essentially an insulator for e.g the La_2CuO_4. The phase diagram is extremely varied depending on the compound. Thus to understand these compounds broadly, we need to focus on the simplest aspects of these cuprates, rather than to invoke more exotic and less generic features. In fig. (1) we exhibit a sketch of the phase diagram for the cuprates. Here, x represents the concentration of holes which can be controlled by adding Sr, for e.g., to $La_{1-x}Sr_xCuO_4$. At zero and small values of doping we obtain an antiferromagnetic phase (AFM) with the ordering temperature $T_N \sim 300K$. Once the AFM order vanishes for a certain value of doping, we observe an insulating phase until a critical concentration above which we observe the insulator-superconductor (SC) transtion. The insulating phase incorporates one of the most controversial topics of today's research – the pseudogap phase. It can be defined as the supression of the spectral weight below a characteristic temperature T^* in the underdoped phase of the compound.

In this region the compounds exhibit a varied range of unsual and anamolous properties. In higher concentration we observe a coexistence of the pseudogap phase and the superconducting phase. Beyond optimal doping (in the overdoped region) we observe the superconducting phase alone. Several theoretical models have been able to give some insight to the most sought after experimental findings of these superconductors. However, we are far from reaching a general consensus about the origin of this phase. The most important

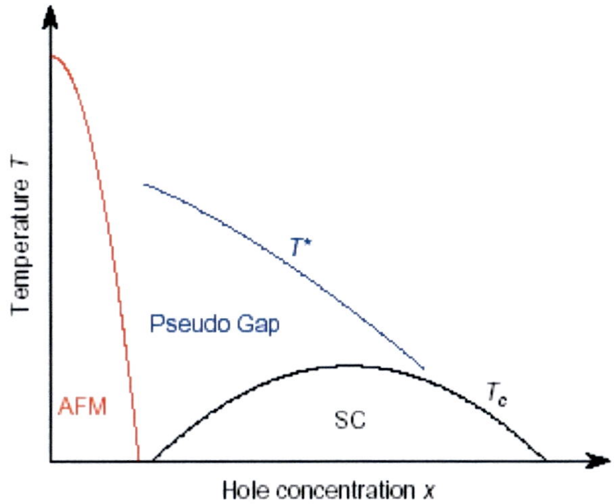

Figure 1. Typical phase diagram of the hole-doped high T_c superconductors.

question about the new high-T_c materials is whether Cooper pairing exists as in conventional superconductors. It is evident from the study of the magnetic flux states of YBCO ring [4] that it exists. Hence, the HTSC are isomorphic with the BCS superconductors. The symmetry sensitive techniques have played a key role in deciding the pairing symmetry of these new materials. They have firmly established a predominant d-wave pairing symmetry. Therefore, unlike the conventional superconductors we assume these compounds to have the $d_{x^2-y^2}$ pairing symmetry (DSC). This type of symmetry with line nodes has been vital in the qualitative change of various properties below T_c. In a simple tight binding model we observe that the penetration depth $\lambda^{-2}(0)/\lambda^{-2}(T)$ approaches unity linearly as T goes to zero [5] and specific heat $C(T)$ is proportional to T^2 [5] for the $d_{x^2-y^2}$ symmetry. The above behaviors are found to be in agreement with the experimental studies [6–8]. The possibility of a mixed angular-momentum-state symmetry has been suggested sometime ago by Ruckenstein et al. [9] and Kotliar [10]. There are also experimental evidences based on Josephson supercurrent for tunneling between Pb and $YBa_2Cu_3O_7$ (YBCO) [11], and on photoemission studies on $Bi_2Sr_2CaCu_2O_{8+x}$ [12] among others which are difficult to reconcile employing a pure s- or d-wave order parameter. These observations suggest that a mixed $[s + \exp(i\theta)d]$ symmetry is applicable in these cases [13, 14]. A theoretical study on the mixed state applying a simple tight binding model has been performed recently in ref. [15]

More recently Krishana et al. [16] have reported a phase transition in the high-T_c superconductor $Bi_2Sr_2CaCu_2O_8$ induced by a magnetic field from a study of the thermal conductivity as a function of temperature and magnetic field. Laughlin [17] has suggested that the new superconducting phase is the time-reversal symmetry breaking $d_{x^2-y^2} + id_{xy}$ state. Similar conclusion has been reached by Ghosh [18] in a recent theoretical study of the temperature dependence of thermal conductivity. This has led to the possibility of the transition to a $d_{x^2-y^2} + id_{xy}$ phase from a pure $d_{x^2-y^2}$ phase of $Bi_2Sr_2CaCu_2O_8$. From a study of vortex in a d-wave superconductor using a self-consistent Bogoliubov-de

Gennes formalism, Franz and Tešanović [19] also predicted the possibility of the creation of a $d_{x^2-y^2} + id_{xy}$ superconducting state. Although, the creation of the mixed superconducting state $d_{x^2-y^2} + id_{xy}$ is speculative, Franz and Tešanović concluded that a dramatic change should be observed (in the observables of the superconductor) as the superconductor undergoes a phase transition from a $d_{x^2-y^2}$ state to a $d_{x^2-y^2} + id_{xy}$ state. A study on the $d_{x^2-y^2} + id_{xy}$ has been performed recently by Ghosh and Adhikari [20] that demonstrates some of the key features of this mixed state superconductivity.

In view of the various experimental evidences and innumerous theoretical studies the current understanding of the pairing symmetry in high temperature superconductors is satisfactory. However, the normal state of these superconductors remains still in darkness and needs a lot of clarification to be understood in full. The article is regarding this enormous task that remains for us to be solved in the underdoped region of the superconductors. In the next subsection we give a short account of the pseudogap based on the various experimental findings. In subsection (1.2) the possible candidates available till date to explain the origin of the pseudogap are introduced. The $d_{x^2-y^2}$-density wave state (DDW) is found to be one of the strong contenders to explain the origin of the pseudogap. So, we would like to introduce the density waves and explain in short the unconventional density waves in section 2. In section 3. we discuss some of the theoretical works on DDW as pseudogap performed on some simple models and compare them with the available experimental measurements. We observe that several experimental findings exhibit a strong accordance with this theory. At the end we conclude our discussion on this topic emphasizing the role of DDW state in the cuprates.

1.1. Pseudogap in HTSC

In this section we give a brief overview of the most important and challenging problems of today's high temperature superconductivity. The depletion of the spectral weight around the fermi level which is more popularly known as the pseudogap is one of the most prominent differences between the conventional supercondutors and the cuprates. It is observed in the underdoped region of the phase diagram of the cuprates and below a characteristic temperature T^* as shown in fig. (1). Various strange phenomena like the presence of the energy gap, anomalous isotope effect, Hall effect etc. are observed in this region above T_c. However, there is a general consensus regarding the temperature and doping region where it exists. Moreover, the $d_{x^2-y^2}$-wave nature of the pseudogap is identical to that of the superconducting gap. In this section, we discuss some of the key experiments which not only provide evidences for the existence of the pseudogap but also reveal important informations about its origin. A detailed report on the experimental evidences can be obtained in the works of Timusk and Statt [21] and Tallon and Loram [22]. The NMR experiments in YBCO were one of the first ones to indicate the existence of a pseudogap. A decrease in the susceptibility in the underdoped phase supports the hypothesis of the gap [23]. The pseudogap can be observed in the Cu nuclei as well as in the O nuclei in the CuO planes of these layered compounds. Many compounds exhibit pseudogap in their magnetic resonnace studies. Some of them are $YBa_2Cu_4O_8$ [24], $Y_2Ba_4Cu_7O_{15}$ [25], $Bi_2Sr_2CaCu_2O_8$ [26] and $HgBa_2Ca_2Cu_3O_{8+\delta}$ [27]. The studies, however, indicate no connection between the presence of the pseudogap and the number of adjacent layers in the compound. The effects on

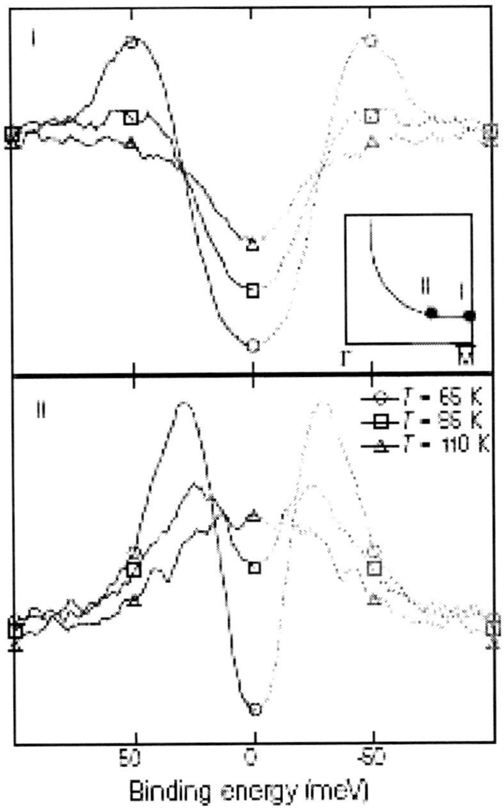

Figure 2. Angle-resolved photoemission [30] on underdoped Bi2212 with T_c=75K at the two points in the Brillouin zone shown in the inset.

superconductivity and on the pseudogap due to the Zn substitution for Cu have revealed some important informations regarding their behavior. Zn has been found to be very effective in suppressing the T_C. However, the NMR Knight shift [28, 29] studies show that the pseudogap remains unaltered by the Zn substitution. This observation and various others have played a key role to show that the superconductivity and pseudogap are independent and competing.

The ARPES (Angle Resolved Photo Emission Spectroscopy) data based on the classical photoelectric effect have been vital not only in detecting the pseudogap but also important in revealing the functional form and its connection with the superconducting gap. The studies on Bi-2201 and Bi-2212 are some that deserve to be mentioned. In fig. (2) we see that the gap energy decreases with increasing doping unlike the superconducting gap which increases [30]. Moreover, it is maximal near $(0, \pi)$ and vanishes along the line connecting (0,0) and (π, π). Remarkably, the pseudogap has the same magnitude as that of the superconducting gap at each k point and evolve continuously into each other as the doping is reduced. But the energetic competition continues to prevail between these two energy scales.

The tunneling experiments using the scanning tunneling microscope (STM) technique

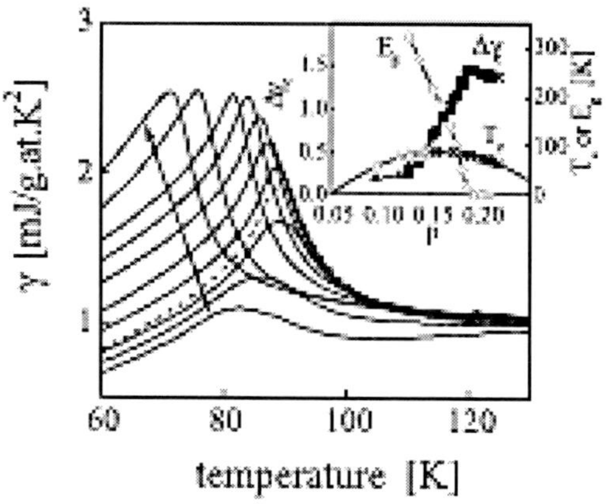

Figure 3. Specific heat γ in mJ/mol K^2 for various doping levels for $Bi_2.15Sr_1.85CaCu_2O_{8+\delta}$.

have been performed very successfully on the Bi2212 crystals [31]. They reveal the pseudogap in the underdoped phase of the cuprates. Moreover, they also demonstrate a close relationship between the superconducting gap as observed by the conductance peak and the pseudogap as exhibited by the depression of the conductivity in the normal state. A smooth evolution through T_c from the superconducting state to the pseudogap state is observed. However, the energy scales are very similar and are temperature independent but depend on the doping level unlike the conventional superconductors.

The electronic specific heat provides an evidence of the gap in the normal state of the cuprates by varying the oxygen doping of YBCO [32] and the Sr content of LBCO [33]. The specific heat coefficient is observed to vary from the underdoped to the overdoped region. We observe a depression in the γ values in the underdoped region below T_c indicating the pseudogap in this region as demonstrated in fig. (3). Similar results have also been found in $La_{2-x}Sr_xCuO_4$ [33] and Bi2212 [34] in the underdoped phase. Moreover, in all the cases we observe that the jump in the specific heat falls sharply at the doping where the pseudogap goes to zero. So we observe that the pseudogap is a generic behavior of cuprates which is evident from the measurement of their thermodynamic property.

Some of the other striking properties in the normal state of the cuprates are the resistivity and Hall effect. In the underdoped materials a deviation from the linear behavior in resistivity (observed in metals) is observed. The behavior not only demonstrates a reduced resistivity but also a break in slope at a temperature correlated with the opening of the pseudogap [35–37]. The deviation from the linear behavior of resistivity can be connected with the appearance of the pseudogap.The above property has been observed in LSCO, YBCO and also in Hg 1223. The temperature dependent Hall coefficient in the normal state of this cuprates is also an important factor for the determination of the pseudogap. In a normal metal, hall coefficient ($R_H = 1/nec$) is a constant and depends on the carrier concentration. It is temperature dependent below T^* and above which it ceases to be temperature independent and thus tends to a constant value [38]. Moreover, the Hall angle θ_H is also found to depend on T, i.e., $\cot(\theta_H) = AT^2 + B$ below the characteristic temperature T^* [39].

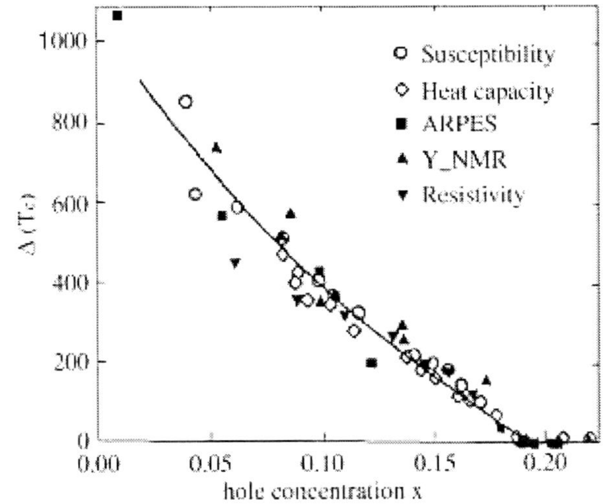

Figure 4. The p-dependence of the pseudogap energy determined from susceptibility, heat capacity, ARPES, ^{8}9Y-NMR and resistivity from [22].

Various other experimental techniques like the Raman scattering [40], optical conductivity [41], spin susceptibility [42] also indicate the presence of pseudogap. The most important aspect which we need to emphasize is the fact that there exists a pseudogap at the onset of superonductivity which vanishes at $x \sim 0.2$ and it has nearly the same size at T_c as can be seen from the measurements of different kinds of probe in fig. (4) [22].

1.2. Some Theoretical Explanations of the Origin of the Pseudogap

Neither the experimental evidences nor the various theoretical models proposed so far have yet been able to lead to a single consistent view point regarding the pseudogap. We are far from reaching a general consensus regarding this effect. In this section, we give a brief overview of some of the theoretical models used so far to study this phenomenon. Due to the time and space constraint it will not be possible to provide a full description of all the models with complete list of references. I hope we will be excused for the partial referencing and citations.

Several models consider the scenario of the preformed pairs to explain the pseudogap in the underdoped phase of the cuprates. One of the simplest ideas has been the BCS-BEC crossover picture [43,44]. It describes a smooth evolution of the system of weakly coupled BCS type to a strongly coupled Bose Einstein condensate of diatomic molecules. Hence, the BCS theory (essential for the formation of the Cooper pairs) is intimately related to BEC (condenstaion of diatomic molecules) and can be explained as the two limits of the same system which can be obtained by varying the chemical potential μ. In this scenario, the pseudogap state has been described as a natural consequence of local, dynamic pairing correlations in the normal state of low-density small pair-size superconductors. They [43, 44] have shown that unlike the BCS case the pairing and the phase coherence of the $q = 0$ pair state takes place at two temperatures. The phase coherence is lost at the critical temperature T_c but the pairs dissociate only at a much higher temperature T^*. In this mean-field scheme where the main regime of interest is in the intermediate coupling regime the gap below T_c is given by

$$\Delta^2(T) = \Delta_{SC}^2(T) + \Delta_{PG}^2(T) \tag{1}$$

where Δ_{PG} is the pseudogap contribution and Δ_{SC} is the superconducting order parameter. Various transport and thermodynamic properties have been successfully calculated with this approach [44].

The pairing of the spinons to form a gap in the spin excitations is shown to be related to the pseudogap. It involves the spin-charge separation [45–47] and was first proposed by P. W. Anderson and is known as the resonating valence bond theory (RVB). Various elaborations of the RVB theory have been made to understand the pseudogap state. The spin charge separation creates holons with zero-spin and spinons with zero-charge spin 1/2 fermions. The Bose condensation of the holons at T_c creates the superconducting gap.

The mechanism suggested by Emery and Kivelson [48] involves the concept of a phase fluctuation of the superconducting order parameter in the pseudogap regime. The local superconducting order parameter $\Delta = \Delta e^{\theta}$ has a well defined amplitude with a fluctuating phase θ. The temperature scale of phase fluctuations is controlled by the zero-temperature superfluid stiffness. Unlike the conventional superconductors, the transition is controlled by phase ordering and explains the Uemura scaling [49]. It is important to mention that in the regime of interest, it is the vanishing of the superfluid density that controls the transition and determines T_c and not the collapse of the gap. Moreover, the reduction of the phase stiffness rapidly drops to zero for conventional s-wave superconductors unlike the $d_{x^2-y^2}$-wave cuprate superconductors. Hence, the temperature range for which we observe the phase stiffness is much wider in the new superconductors and can be observed at temperatures far from T_c.

The nearly antiferromagnetic Fermi liquid (NAFL) model to describe this phenomenon has been proposed by Pines and collaborators [50]. The close proximity to the antiferromagnetic phase leads to strong anisotropic magnetic interaction which explains the anamolous normal state of the cuprates. The pseudogap in the quasiparticle spectrum is the formation of a precursor to a spin-density wave and is based on the scenario of the dynamic spin susceptibility $\chi(q, \omega)$ which is peaked at the commensurate wave vector $Q \approx (\pi, \pi)$. Due to the highy anisotropic behavior of the effective interation, two classes of quasiparticles on the Fermi surface is defined: hot quasiparticles in the vicinity of the hot spots where the magnetic interaction is anomalously large and the cold quasiparticles where the interaction is normal and behave like the Landau-Fermi liquid. A gap in the hot quasiparticle spectrum below T^* is identified as the pseudogap whereas the gap in the cold quasiparticle spectrum below T_c establishes the superconducting gap.

There have been various other models to study this feature in the underdoped cuprates like the interlayer exchange coupling [51] for bilayer compounds, the appearance of a quantum critical point [52] at optimal doping associated with a charge density wave instability etc. Without going to much details regarding these models we would like to discuss in the next section the d_{x-y^2}-density wave scenario (DDW) where this new order parameter is one of the most promising competing models for this strange phenomena.

2. Density Waves and HTSC

Density waves are formed due to the interaction between the electrons in a metal. Due to the distortion of the lattice intermediating the two electrons (electron-electron interaction) we often find the pairing of electrons. These instabilities lead to the Cooper pairing on one hand and charge density waves (CDW) on the other hand. The ordering of electrons below a characteristic temperature, known as the charge-density waves, were envisaged by theoretical physicist Rudolg E Peierls in 1930s and was finally observed in 1970. The characteristics of the CDW and Cooper pairing are very similar. That's why at one time CDW's were suggested as being the cause of superconductivity. This self organization into regular patterns leads to a very high dielectric constant which allows them to store enormous charge. Moreover, CDW behaviors vary distinctly and peculiarly to the applied voltages which can be direct or alternating or a sum of both. A report on the density waves is available in the article of Brown and Gruner [53]. CDW has been extensively studied in the context of superconductivity to explain the semiconducting phases of the $BaPb_{1-x}Bi_xO_3$ or $Ba_{1-x}K_xBiO_3$ [54] which can be attained by varying the doping. The pure CDW phase, SC phase or the mixed CDW+SC phase have been observed in these compounds. These compounds were also found to have T_c much higher than the other oxides. CDW supercondutors include layered dichalcogenides, $NbSe_3$, some organic substances and compounds with A15 and C15 structures. A much detailed review on these substances can be obtained from ref. [55]. The interpretation of the pseudogap of high temperature superconductors in terms of the CDW [56] has renewed the interest in the study of this order for these new materials. The cuprates involve a order parameter competition which can be observed in the characteristics of the transition temperature T_c. It initially grows and then decreases as we vary the doping. Moreover, the pseudogap is a phase with broken symmetry in the $d_{x^2-y^2}$ state which leads to the proposal of a d-density wave or an unconventional density wave parameter. A short discussion on these density waves are given below.

2.1. Unconventional Density Waves

The DDW phase could be a result of the competition between an effective attraction and the short range repulsion which leads to the formation of the superconducting states. The order parameter of a singlet DDW state given by $W_{\mathbf{Q}}$ can be defined as

$$W_{\mathbf{Q}} f_k = \langle c^\dagger_{\mathbf{k}'\uparrow} c_{\mathbf{k}'+\mathbf{Q}\uparrow} + c^\dagger_{\mathbf{k}'\downarrow} c_{\mathbf{k}'+\mathbf{Q}\downarrow} \rangle \tag{2}$$

where f_k is the form factor which refers to the angular momentum state of the parameter and is given by $\cos k_x a - \cos k_y a$. This order parameter breaks the time reversal symmetry, parity, translational symmetry and rotational symmetry by $\pi/2$. But under an action of any two we observe that the symmetry is preserved. This also carries a finite bond current.

The commensurate $Q = (\pi/a, \pi/a)$ singlet $d_{x^2-y^2}$-density wave state is also the staggered flux state, as we can see that the flux in each plaquette is alternating. We can also have an incommensurate density wave state [57]. $W_{\mathbf{Q}}$ must be imaginary when we have $f(k+Q) = -f^*(k)$ which is the case for DDW in high temperature superconductors. However for the incommensurate ordering we can have $W_{\mathbf{Q}}$ to have arbitrary phase. It also

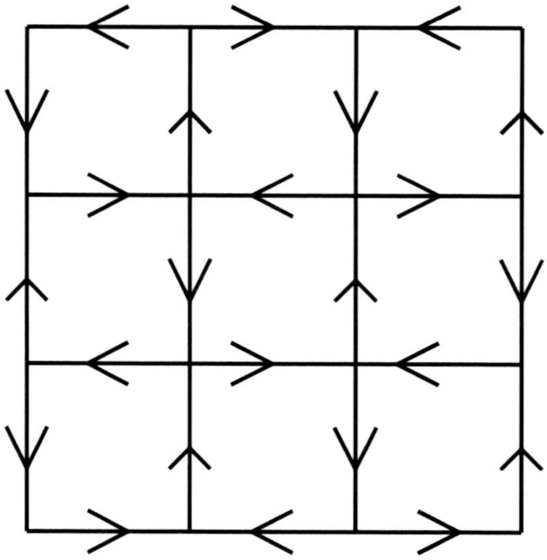

Figure 5. $d_{x^2-y^2}$ density-wave state.

breaks the translational and rotational symmetries. The time reversal symmetry is also broken unless $W_{\mathbf{Q}} = W^*_{-\mathbf{Q}}$. Moreover the DDW transition is in the same universality class as the random-bond Ising model. Here we have concentrated only on the singlet-density waves ie. $d_{x^2-y^2}$ waves as proposed by various experiments on pseudogap. A much detailed study on the unconventional density waves can be found from the works of Nayak [58] where it is shown that unlike the d-wave state the p-wave state leads to the Peierls ordering.

3. Pseudogap and DDW Order

In this section we argue that the features of the pseudogap may be linked to the development of a new order parameter, i.e., the d-density wave order, identified by the parameter that explains the physics of the staggered flux state as stated above. In the beginning we would like to explain the phase diagram encountered in the cuprates using simple models within the mean field approximation and study the extent to which it is capable of explaining the essence of our idea. The idea of modelling the pseudogap using DDW was proposed by Chakrabarty et. al. [56]. This gap, according to this proposal, is a real gap in the one-particle excitation spectrum at the wave vector (π,0). However, it is "pseudo" in experiments because of the sample imperfection and other experimental difficulties. The sample preparation difficulties not only lead to complications in detection but also lead to anamolous experimental results in the underdoped phase. This results in a divergence of opinion and is thereby delaying the process of reaching a general consensus as to the origin of pseudogap. In an attempt to understand this strange phase, we would like to first study the phase diagram and compare it with that in fig (1). The next step would be to calculate various properties (thermodynamic, electromagnetic etc.) and compare them with the experimental measurements.

3.1. Phase Diagram

In this section simple models within the mean field scenario are considered to describe the basic aspects of the phase diagram of the cuprates. In the first place a simple two-dimensional tight binding model with appropriate lattice symmetry with second nearest hopping parameter is considered. A mean-field calculation is performed to obtain the phase diagrams with doping, δ [59]. The model could be understood as an extension of the Hubbard model with nearest-neighbor interaction and on-site interaction. In the beginning both the spin density waves and the charge density waves are considered to determine which one of them is more appropriate to describe the phase diagram of fig. (1).

The Hamiltonian can be written as

$$H = \sum_{\mathbf{k}\sigma}(\epsilon_{\mathbf{k}} - \mu)c^{\dagger}_{\mathbf{k}\sigma}c_{\mathbf{k}\sigma} + \sum_{\mathbf{k}\sigma} W_{\mathbf{k}}c^{\dagger}_{\mathbf{k}\sigma}c_{\mathbf{k}+\mathbf{Q}\sigma} + h.c + \sum_{\mathbf{k}} \Delta_{\mathbf{k}}c^{\dagger}_{\mathbf{k}\uparrow}c^{\dagger}_{\mathbf{k}\downarrow} + h.c \qquad (3)$$

where $c^{\dagger}_{\mathbf{k}\sigma}$ is the fermion creation operator for wavevector $\mathbf{k}$ and spin σ. $W_{\mathbf{k}}$ and $\Delta_{\mathbf{k}}$ are the density wave gap and superconducting gap respectively. μ is the chemical potential. At half-filling $\mu = 0$ with nearest neighbor hopping, while it takes negative values away from half filling. The density wave gaps are given by

$$\begin{aligned} W_{\mathbf{k}} &= V_{DW}\phi_{\mathbf{k}}\sum_{\mathbf{k}'}\langle c^{\dagger}_{\mathbf{k}'\uparrow}c_{\mathbf{k}'+\mathbf{Q}\uparrow} - c^{\dagger}_{\mathbf{k}'\downarrow}c_{\mathbf{k}'+\mathbf{Q}\downarrow}\rangle \quad \text{(SDW)} \\ W_{\mathbf{k}} &= V_{DW}\phi_{\mathbf{k}}\sum_{\mathbf{k}'}\langle c^{\dagger}_{\mathbf{k}'\uparrow}c_{\mathbf{k}'+\mathbf{Q}\uparrow} + c^{\dagger}_{\mathbf{k}'\downarrow}c_{\mathbf{k}'+\mathbf{Q}\downarrow}\rangle \quad \text{(CDW)} \end{aligned} \qquad (4)$$

where CDW (SDW) represents the charge (spin) density state, $\mathbf{Q}$ is the nesting vector (π, π) and V_{DW} is the pairing interaction for the density wave channel. $\phi_{\mathbf{k}}$ represents the basis function which determines the symmetry of the order parameter and the corresponding pairing interaction. For the conventional SDW or CDW, $\phi_{\mathbf{k}} = 1$ and for d-density waves, $\phi_{\mathbf{k}} = \cos k_x - \beta\cos k_y$. Here $\beta = 1$ corresponds to square lattice and $\beta \neq 1$ represents orthorhombic distortion. The superconducting gap is of $d_{x^2-y^2}$ wave type (DSC) and can be defined as

$$\Delta_{\mathbf{k}} = V_{DSC}\eta_{\mathbf{k}}\sum_{\mathbf{l}}\langle c_{-\mathbf{l}\uparrow}c_{\mathbf{l}\downarrow}\rangle \qquad (5)$$

where V_{DSC} is the pairing interaction and $\eta_{\mathbf{k}}$ is the basis function that determines the symmetry of the superconducting order parameter . The energy dispersion relation is given by $\epsilon_{\mathbf{k}} = -2t(\cos k_x + \beta\cos k_y - 2\gamma\cos k_x\cos k_y)$ which gives the band structure. t and βt are the nearest neighbor hopping integrals along the in-plane a and b axes, respectively. The model is more realistic due to the next-nearest neighbor hopping γt. The order parameter of the density wave is

$$W_{\mathbf{k}} = iW_0(\cos k_x - \beta\cos k_y) \qquad (d_{x^2-y^2}\ \text{wave})$$

$$W_{\mathbf{k}} = W_0 \qquad (s\ \text{wave}).$$

The d-wave superconducting gap is $\Delta_{\mathbf{k}} = \Delta_0(\cos k_x - \beta\cos k_y)$. The two mean field equations obtained from the stationary conditions are

$$\Delta_0 = \frac{V_{DSC}}{2}\sum_{\mathbf{k}}\sum_{\nu=\pm}\left[\Delta_0\eta^2_{\mathbf{k}} + \alpha\nu\frac{2\Delta_0\eta^2_{\mathbf{k}}W_0^2\phi^2_{\mathbf{k}}}{f(\mathbf{k})}\right]\frac{1}{E_\nu(\mathbf{k})}\tanh\frac{E_\nu(\mathbf{k})}{2T} \qquad (6)$$

and

$$W_0 = \frac{V_{DW}}{2}\sum_{\mathbf{k}}\sum_{\nu=\pm}\left[W_0\phi_{\mathbf{k}}^2 + \nu\frac{W_0\phi_{\mathbf{k}}^2}{f(\mathbf{k})}\left((\xi_{\mathbf{k}} + \xi_{\mathbf{k+Q}})^2 + 4\alpha\Delta_0^2\eta_{\mathbf{k}}^2\right)\right] \times \frac{1}{E_\nu(\mathbf{k})}\tanh\frac{E_\nu(\mathbf{k})}{2T} \tag{7}$$

where α depends on $\phi_{\mathbf{k}}$. $\alpha = 0$ for SDW and $\alpha = 1$ for d-SDW and $\xi_{\mathbf{k}} = \epsilon_{\mathbf{k}} - \mu$. In order to study the doping dependence ($\delta = 1 - n$) of the order parameters it is necessary to consider the number equation given by

$$n = 1 - \frac{1}{2}\sum_{\mathbf{k}}\sum_{\nu=\pm}\left[(\xi_{\mathbf{k}} + \xi_{\mathbf{k+Q}}) + \nu\frac{(\xi_{\mathbf{k}} + \xi_{\mathbf{k+Q}})}{2f(\mathbf{k})} \times \left((\xi_{\mathbf{k}} - \xi_{\mathbf{k+Q}})^2 + 2W_0^2\phi_{\mathbf{k}}^2\right)\right]\frac{1}{E_\nu(\mathbf{k})}\tanh\frac{E_\nu(\mathbf{k})}{2T} \tag{8}$$

The above equations (6), (7) and (8) can be numerically solved for the conventional SDW and d-SDW case. However, a region of coexistence of d-SDW and the superconducting state is not observed. It is only possible to obtain the d-SDW or the SC gap with doping. Hence our model confirms that the d-density wave scenario of the pseudogap is appropriate for the CDW instead of the SDW case. The phase diagrams of DDW can be obtained using the following equations

$$\Delta_0 = \frac{V_{DSC}}{2}\sum_{\mathbf{k}}\sum_{\nu=\pm}\Delta_0\eta_{\mathbf{k}}^2\frac{1}{E_\nu(\mathbf{k})}\tanh\frac{E_\nu(\mathbf{k})}{2T} \tag{9}$$

and

$$W_0 = \frac{V_{DW}}{2}\sum_{\mathbf{k}}\sum_{\nu=\pm}\left\{W_0\phi_{\mathbf{k}}^2 + \nu\frac{W_0\phi_{\mathbf{k}}^2}{g(\mathbf{k})}(\xi_{\mathbf{k}} + \xi_{\mathbf{k+Q}})^2\right\}\frac{1}{E_\nu(\mathbf{k})}\tanh\frac{E_\nu(\mathbf{k})}{2T} \tag{10}$$

In fig. (6) the order parameters Δ_0 and W_0 versus doping phase diagram is plotted for the SDW case . In the region of doping close to zero a dominant density-wave state is observed, while away from $\delta = 0.0$ a dominant superconducting state exists. A coexistence of the above two types of order parameters is found in the region $0 \leq \delta \leq 0.06$. The SDW state extends down to half filling as no measures have been taken to observe an insulating phase at $\delta = 0$. A temperature T versus doping phase diagram for $\gamma = 0.0$ is shown in the inset of the figure. A qualitatively similar dependence of T on δ is observed. It shows a $(T_c)_{max}$ at $\delta = 0.06$ for $\gamma = 0.0$.

The DDW and the DSC states are also studied with doping. Fig. (7) displays the dependence of the order parameters on δ. We consider two sets of values:(i) $V_{DW} = 2.6$ and $V_{SC} = 2.2$ and (ii) $V_{DW} = 2.6$ and $V_{SC} = 2.28$ for $\gamma = 0$. The phase diagram also shows a region of co-existence that depends on the values of V_{DSC} and V_{DW}. Δ_0 increases with δ in the presence of the pseudogap or DDW, reaches a maximum and then decreases rapidly and goes to zero (not shown in the figure). As the pseudogap is believed to have a d-wave symmetry, fig. (7) closely resembles the phase diagram of the cuprates. The order parameters are very sensitive to the value of μ. The relative values of the superconducting

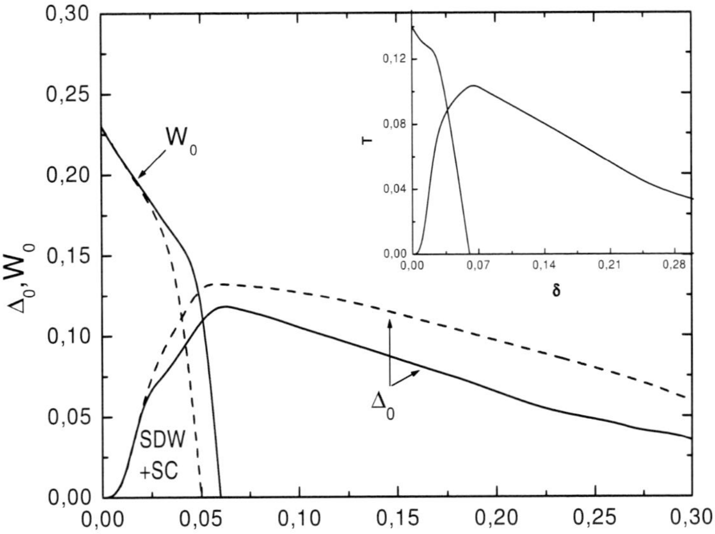

Figure 6. The order parameter vs doping phase diagram for $\gamma = 0.0$ (solid line) and $\gamma = 0.05$ (dashed line) for SDW. Inset shows the temperature vs doping for $\gamma = 0.0$.

gap or the density wave gap can be altered by changing the values of V_{DW} and V_{DSC} as is evident from the figure. In the inset of fig. (7) the plot of T with δ demonstrates qualitatively similar dependence on δ.

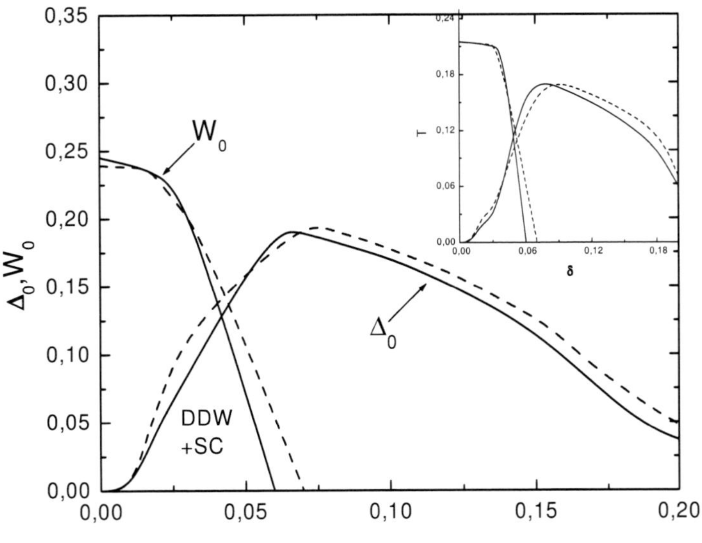

Figure 7. The order parameter vs doping phase diagram for sets (i) solid line and (ii) dashed line for DDW. Inset shows the temperature vs doping for the same sets as above

In fig. (8) the temperature dependencies of DDW and DSC waves are considered for $\delta = 0.01$ for sets (i) and (ii) where both the states co-exist. The dependencies are displayed in two separate boxes to visualize the plots better. In this case also a suppression of the

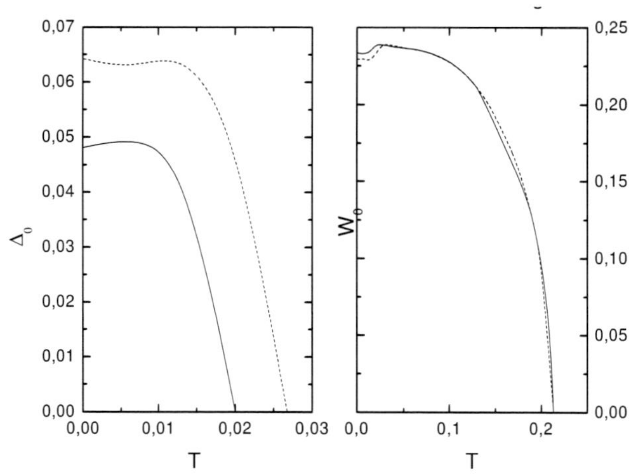

Figure 8. The temperature dependence of DDW and DSC states for $\delta = 0.01$ for sets (i) solid line and (ii) dotted line.

DDW state in the presence of DSC state is observed. Moreover the value of the T_c in the presence of the DDW state is much lower when compared to SDW [59]. This nature can also be observed in the inset of fig. 7. Hence the competition between the DDW and DSC is much higher when compared to the case of SDW

So, in the above study both the conventional and unconventional density waves have been considered. The gap equations obtained for SDW, d-SDW and DDW are all different. d-SDW wave does not give any solution for pseudogap in the presence of DSC state. So if we believe the pseudogap to have a $d_{x^2-y^2}$ symmetry, the charge density wave or rather the DDW state is more appropriate where a dominant density wave is observed at doping close to half filling. With increase of doping the DSC gap coexists with the density wave and then we find the DSC gap alone. The phase diagram is in complete agreement with the experimental preditions.

To seek further insight on the phase diagrams of the cuprates it is of utmost importance to study a phase diagram which includes the anti-ferromagnetic phase at half-filling which was not considered in the above model. A generalization of the Hubbard model which includes next-neighbor repulsion and hopping was able to incorporate the AF phase, DSC phase and also the DDW phase [60]. The final form of the reduced Hamiltonian is

$$\begin{aligned} H &= \sum_{\mathbf{k}\sigma\lambda}(\epsilon_{\lambda\mathbf{k}} - \mu_\lambda)c^\dagger_{\lambda\mathbf{k}\sigma}c_{\lambda\mathbf{k}\sigma} - V_{AF}\sum_{\mathbf{k}\mathbf{k}'} c^\dagger_{\mathbf{k}+\mathbf{Q}\sigma\lambda}c_{\mathbf{k}\sigma\lambda}c^\dagger_{\mathbf{k}'\sigma'\lambda}c_{\mathbf{k}'+\mathbf{Q}\sigma'\lambda} \\ &- V_{DDW}\sum_{\mathbf{k}\mathbf{k}'} g(\mathbf{k})g(\mathbf{k}')c^\dagger_{\mathbf{k}+\mathbf{Q}\sigma\lambda}c_{\mathbf{k}\sigma\lambda}c^\dagger_{\mathbf{k}'\sigma'\lambda}c_{\mathbf{k}'+\mathbf{Q}\sigma'\lambda} \\ &- V_{DSC}\sum_{\mathbf{k}\mathbf{k}'} g(\mathbf{k})g(\mathbf{k}')c^\dagger_{\mathbf{k}\uparrow\lambda}c^\dagger_{-\mathbf{k}\downarrow\lambda}c_{\mathbf{k}'\uparrow\lambda}c_{-\mathbf{k}'\downarrow\lambda} \end{aligned} \tag{11}$$

where ϵ_{k11}=ϵ_{k11}=ϵ_k=$-2t(\cos k_x + \beta\cos k_y - 2\gamma\cos k_x\cos k_y)$ and $\epsilon_{k12} = \epsilon_{k21} = (t_\perp/4)f(k)^2$

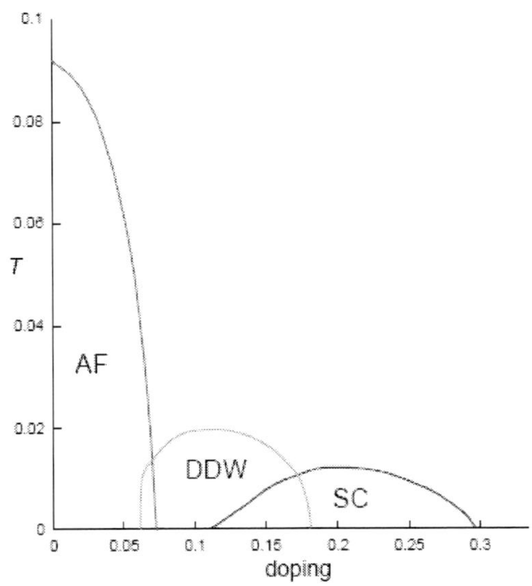

Figure 9. Phase diagram for $t = 0.5$ eV, $\gamma t = -0.025$, $t_{\perp} = 0.05$ eV, $g_{AF} = 0.06$ eV, $g_{DDW} = 0.02$ eV, $g_{DSC} = 0.01$ eV.

The mean-field phase diagram of this model in fig(9) can be obtained by minimizing the free energy expression of the above Hamiltonian which can be observed to incorporate the three important phases of the HTSC as shown in fig. (1).

We observe regions of coexistence namely the DDW+AF and also DSC+DDW states. The regions of existence of the three phases can be altered by varying the parameters of this model. So the above two models successfully describe the insulating, pseudogap and the superconducting phases of the cuprates. Fig (9) is a typical illustration of the phase diagram obtained from the above model.

Calculations on the $t - J$ model including the flux phases could satisfactorily explain the pseudogap phase. The flux phases in the $t - J$ model is identical to the DDW phase as we have seen in our earlier discussions. In ref. [61] the phase diagram of the t-J model was studied using a mean-field type approximation within the Baym-Kadanoff expansion for Hubbard X operators. The generalized $t - J$ Hamiltonian in this case can be written as

$$\begin{aligned} H &= -\sum_{ijp=1..N} \frac{t_{ij}}{N} X_i^{p0} X_j^{0p} + \sum_{ijp,q=1..N} \frac{J_{ij}}{4N} X_i^{pq} X_j^{qp} \\ &- \sum_{ijp,q=1..N} \frac{J_{ij}}{4N} X_i^{pp} X_j^{qq} + \sum_{ijp,q=1..N} \frac{V_{ij}}{2N} X_i^{pp} X_j^{qq} \end{aligned} \quad (12)$$

where i and j run over the sites of a lattice. t_{ij} and J_{ij} are the hopping and exchange constants, respectively. Both the above interactions are limited to nearest neighbors only. V_{ij} describes the Coulomb repulsion between the electrons. The phase diagram obtained from this calculation also shows the same characteristics of the DDW and DSC phases as can be seen from the figure (10).. The phase diagram clearly depicts the characteristics of the phases of the cuprates.

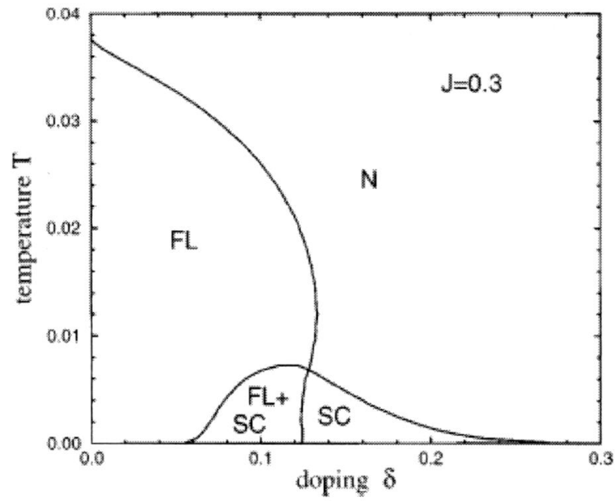

Figure 10. Phase diagram of the $t - J$ model.

Hence, we see that the pseudogap state can be understood using the DDW phase. Moreover, the competition between the DDW and the DSC phases is clearly visible form the phase diagram which demonstrates the exact nature of the pseudogap and superconducting gap.

3.2. Pseudogap and Its Various Properties

In this section we would like to discuss the various thermodynamic, electromagnetic and transport properties of the high temperature superconductors which could successfully be explained by the DDW theory of the pseudogap and could help us to decipher the mystery of this state. It has been observed experimentally that the jump of the specific heat data at T_c depends on the doping level [62]. The specific heat can be calculated from the two-dimensional model given by eq. (3). The temperature dependence of specific heat $C_s(T)$ can be written as $C_s(T) = T(dS/dT)$ where the entropy S is

$$S(T) = -2K_B \sum_{\mathbf{k}} [(1 - f_{\mathbf{k}}) \ln(1 - f_{\mathbf{k}}) + f_{\mathbf{k}} \ln f_{\mathbf{k}}] \tag{13}$$

and $f_{\mathbf{k}}$ is the Fermi function.

In fig. (11) the temperature dependence of the ratio of the superconducting specific heat and the normal specific heat, $C(T)/C_n(T_c)$ at T_c is displayed for various values of δ. The values of doping are chosen from the phase diagram given in fig. (2). The jump at T_c increases as δ decreases. The jump is maximum for $\delta = 0.7$ where we have only the superconducting state. This nature with doping has also been predicted by experiments [62]. The plots clearly show the presence of a pseudogap above T_c.

The spin and current correlation function in the DSC state, DDW state and the mixed DSC+DDW state are found to strengthen the candidature of the DDW state as the pseudogap in the underdoped regime of superconductors [63]. The Hartree-Fock and RPA theories are employed to identify any difference in the behavior of the three phases. Irrespective of

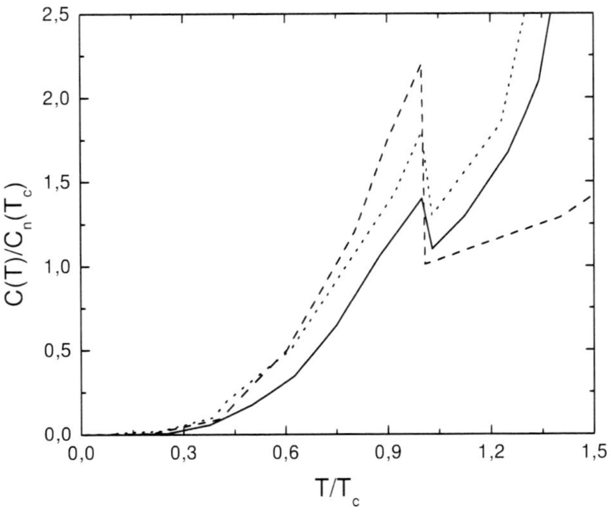

Figure 11. Temperature dependence of the specific heat for various doping concentrations for $\delta = 0.02$ (solid line), 0.04 (dotted line) and 0.3 (dashed line).

the shortcomings of the above model close to the quantum critical point the theory illuminates the understanding of the material in the underdoped regime. It is evident from the fig. (12) that the resonant peak in the underdoped regime is due to the existence of the DDW phase. Both below and above T_c the signature stays intact. The dependence of superfluid density on doping and temperature shows behavior similar to that observed in experiments. The DDW order weakens with the increase in x and leads to the increase in the superfluid density as shown in fig. (13). The change of the linear dependence of superfluid density on temperature in the optimal and moderately doped samples to the $\sqrt{(T)}$ behavior in the heavily doped samples or in the intermediate temperature range in fig. (13) clearly indicates the presence of the DDW state.

The electronic structure in the vicinity of impurities can be used to distinguish the DDW and the DSC phases. The study of the local DOS (density of states) in the DDW state and DSC states in the presence of of a non-magnetic impurity [64] gives a qualitative difference in the nearest-neighbor and the next-nearest neighbor sites. The results of the clean case and the DOS on the impurity site fail to indicate any qualitative difference between the two. The non-magnetic impurity case can be described by the Hamiltonian

$$H_{sc} = U_0 \sum_{k,k',\sigma}' \psi^{\dagger}_{k,\sigma} \hat{\tau} \psi_{k',\sigma}, \tag{14}$$

where U_0 is the scattering strength of the impurity, $\psi^{\dagger}_{k,\sigma}$ is the usual spinor in the DDW state. The dispersion relation is the same as that used in the ref [59]. The electronic Greens function is given by $\hat{G}_{\sigma\sigma}(\mathbf{k}, \tau - \tau') = - < T\psi_{k,\sigma}(\tau)\psi^{\dagger}_{k,\sigma}(\tau') >$. In the presence of the scattering potential as in eq. (14) the Greens function is given by

$$\hat{G}(r, r', i\omega_n) = \hat{G}_0(r' - r, i\omega_n) + \hat{G}_0(-r, i\omega_n)\hat{T}(i\omega_n)\hat{G}_0(r', i\omega_n) \tag{15}$$

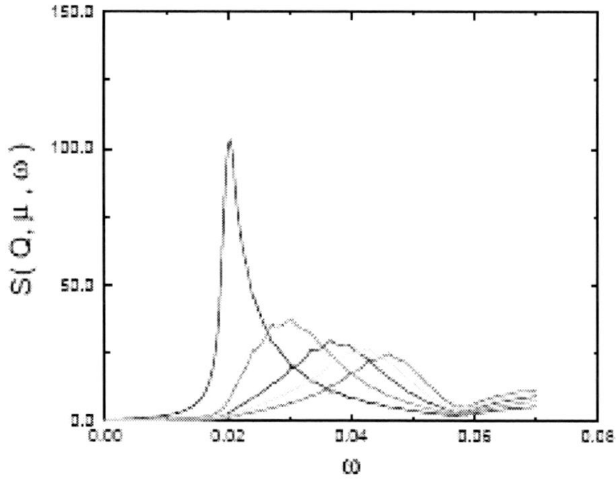

Figure 12. Spin-spin correlation function as a function of ω for various reduced Coulomb repulsion with coexisting singlet DDW and DSC orders with $t = 0.5$ eV, $\Delta_0 = .02$ eV, $W_0 = .02$eV and μ set to -.01 eV.

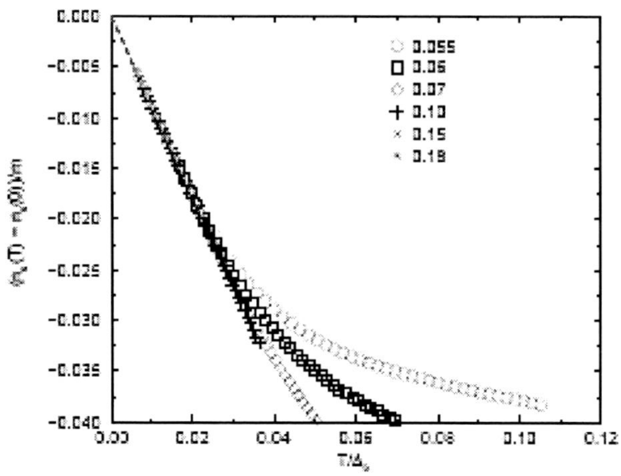

Figure 13. The temperature-dependent part of the superfluid stiffness plotted against scaled temperature for six values of the doping. The doping concentrations are indicated in the legends. The values of W_0 and Δ_0 for a given doping are derived from Eqs. 15 and 16. μ is derived from $\mu = -x^2$ and t = 0.5 eV. The dotted line is an extrapolation of the results to zero temperature

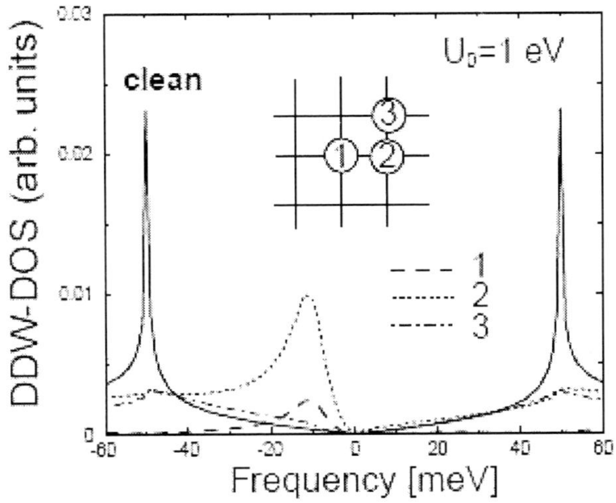

Figure 14. DDW-DOS for the clean case (solid line) and in the presence of a non-magnetic impurity. (1) DOS on the impurity site (2) DOS on the nearest neighbor site (3) DOS on the next-neighbor site

where $\hat{T}(i\omega_n)$ is the T-matrix. The DDW-DOS can be calculated numerically as

$$N(\omega) = -\frac{2}{\pi}\mathrm{Im}\hat{G}^{11}(r, r, \omega + i\delta) \tag{16}$$

The superconducting case can similarly be calculated and the results are plotted in the following figs (14) and (15).

Similarly we can consider the magnetic impurity which induces two resonnace states in the DDW-DOS, a result similar to the DSC state. Hence, the results seem to be of interest in the detection of the DDW order.

More recently, the study of the single impurity scattering on the local density of states has also been performed on the DSC state, DDW state and the coexisting DSC+DDW state [65]. In this case the Hamiltonian given by eq. (3) along with the impurity scattering term given by

$$H_{imp} = \sum_{k,k'}\sum_{\sigma\sigma'=\uparrow,\downarrow} V_{k,k'} c^{\dagger}_{k\sigma} c_{k'\sigma'} \tag{17}$$

is used. Using the T-matrix formalism similar to the work above it is possible to study the above mentioned phases in the presence of the magnetic-impurity scattering and also the non-magnetic scattering. In the pure DSC state the features are standard and very similar to the results observed in ref. [66] but indicate a clear dependence on the band structure when compared with that of ref. [66]. In the case of the mixed DSC+DDW phase we observe a clear signature of the DDW state when the energies are higher than the superconducting gap. Below the DSC gap energy we have the DSC state to dominate. A sudden change is observed at the value of the gap. In the pure DDW wave we observe the peaks to be more broadly spaced and very sensistive to the parameter changes. The STM experiments should thus be very useful in the determination of the origin of the pseudogap as can be seen from fig. (16).

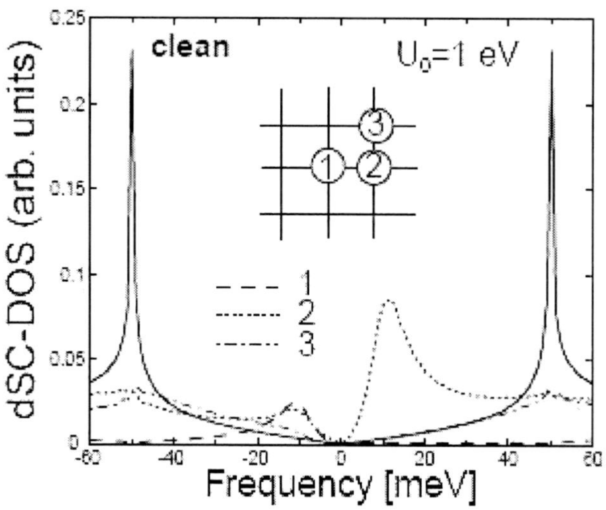

Figure 15. DSC-DOS for the clean case (solid line) and in the presence of a non-magnetic impurity. (1) DOS on the impurity site (2) DOS on the nearest neighbor site (3) DOS on the next-neighbor site

In ref. [67] the DDW order is considered using a more realistic band structure consistent with the angle resolved photoemisssion spectroscopy (ARPES) data. The disperion of the wavevectors with energy in fig. (17) is observed to be similar to the experimental measurements whereas the dispersion with energy is larger. Moreover, it is also sensitive to the choice of the band structure. So, the agreement between the calculation and experiment is an indication of the DDW phase. The quantitative difference could be due to electron-electron interaction which becomes important for the comparison. The presence of two non-magnetic impurities and the quantum interference between the two have also been studied in the scenario of DDW state [68]. It could also act as a very useful tool to decipher the controversy of the issue.

The effect of the non-magnetic impurities has also been studied in the two-dimensional $t-J$ type model using the Bogoliubov-de Gennes (BdG) method [69] to study the effect of the impurities on the DDW and DSC order. From fig. (18) we see that the superconducting gap is severely affected by disorder whereas the DDW order is robust to the above effect. Differences in the behavior of the averaged DOS are observed where we see that with the increasing impurity the DSC state ceases to exist. However, the DDW order persists. So the experiments based on these theoretical predictions could be of great importance for the problem.

The effect of the magnetic field [70–72] on the DDW states, the DSC states as well as the coexisting DDW+DSC states have been considered. Let us first consider the effect of a magnetic field applied perpendicular to the CuO planes on the DDW order. The mean-field hamiltonian of DDW under the magnetic field with the choice of the Landau gauge [70] is

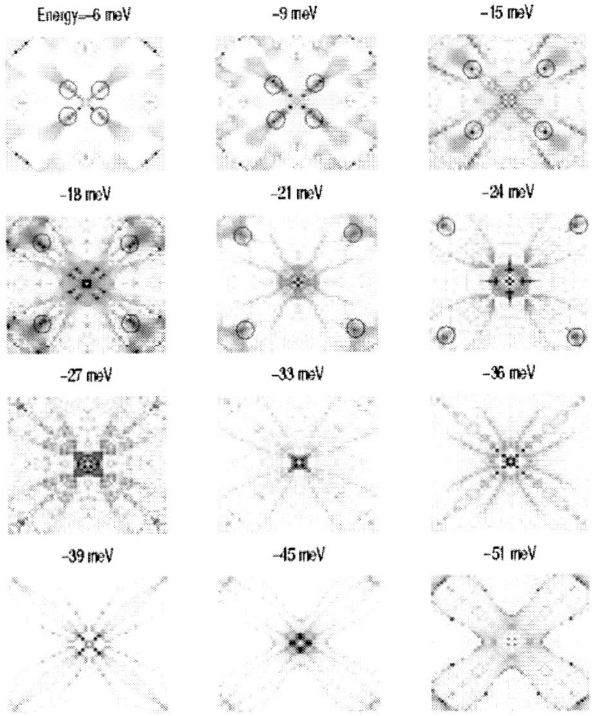

Figure 16. Fourier transform of STM in the coexisting DSC-DDW state with the $t - t'$ - band structure, DSC gap of 25meV and DDW gap of 40meV, for nonmagnetic impurity scattering of 100meV. The peaks along the directions (π, π) are marked by circles. As noted, these peaks get close to the corners of the BZ at energies equal to the DSC gap, and not to the total gap, which is 47meV. Also, the general features of the spectra change dramatically at energies larger than DSC gap.

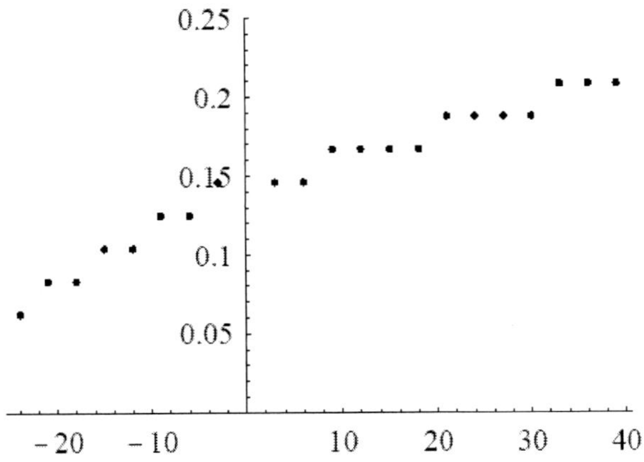

Figure 17. The energy dispersion of the wavevectors of the observed peaks. The magnitude of the wavevectors is plotted in units of 2π and the energy is measured in meV. The magnitude of DDW gap is 40 meV

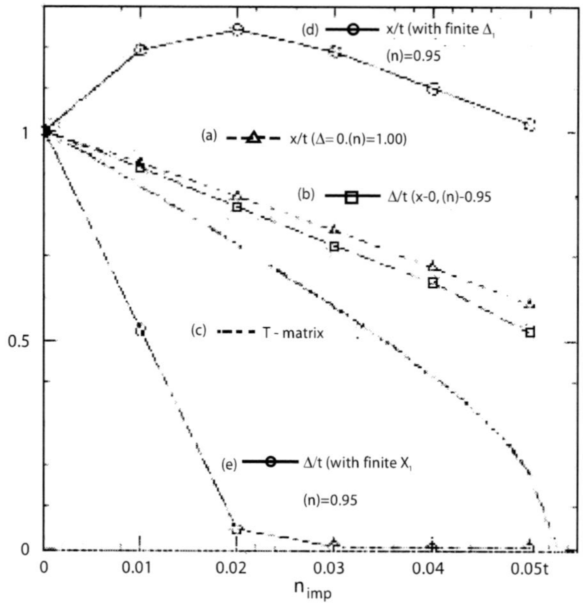

Figure 18. The evolution of DSC gap and DDW gap with n_{imp}.

given by

$$
\begin{aligned}
H &= -\mu \sum_{m,n} c^{\dagger}_{m,n} c_{m,n} - t \sum_{m,n} (c^{\dagger}_{m,n} c_{m+1,n} + h.c) - t \sum_{m,n} (e^{2\pi i \alpha m} c^{\dagger}_{m,n} c_{m,n+1} + h.c) \\
&- \sum_{m,n} (-1)^{m+n} \Delta (-i c^{\dagger}_{m,n} c_{m+1,n} + h.c) + \sum_{m,n} (-1)^{m+n} \Delta (-i e^{2\pi i \alpha m} c^{\dagger}_{m,n} c_{m,n+1} \\
&+ h.c)
\end{aligned} \tag{18}
$$

where m and n are the site labels of the square lattice in the x and y directions; $\alpha = \phi/\phi_0$ is the magnetic flux through each plaquette of the square lattice in terms of the flux quantum $\phi_0 = hc/e$ and $c_{m,n}$ is the fermion destruction operator at the site m, n. The Fourier transformed Hamiltonian can be written to calculate the DDW gap magnitude with the applied magnetic field as can be seen in fig. (19).The DDW gap can be written as

$$\Delta = i(-1)^{m+n} V < e^{2\pi i \alpha m} c^{\dagger}_{m,n} c_{m,n+1} - h.c > \tag{19}$$

We notice from the above study that the effect of the magnetic field on the DDW order is rather weak when compared to the DSC state as is supported by the nuclear-magnetic resonance experiments in $YBa_2Cu_4O_8$ [73] and also $YBa_2Cu_3O_{7-\delta}$ [74]. Moreover, unlike the DSC case, we observe formation of the Landau levels in the DDW case and thus indicates a possible formation of the de Haas-van-Alphen (dHvA) oscillations in the pseudogap phase. The dHvA oscillations is plotted in fig. (20) which indicate the magnetization oscillations to be very weak making it difficult for observations in experiments.

The Nernst effect in the pseudogap phase of the cuprates has also been studied due to the magnetic and electric fields. The electric and magnetic fields can be parallel and perpendicular to the conducting plane [71]. The Nernst coefficient is given by $S_{xy} = -S/B\rho$

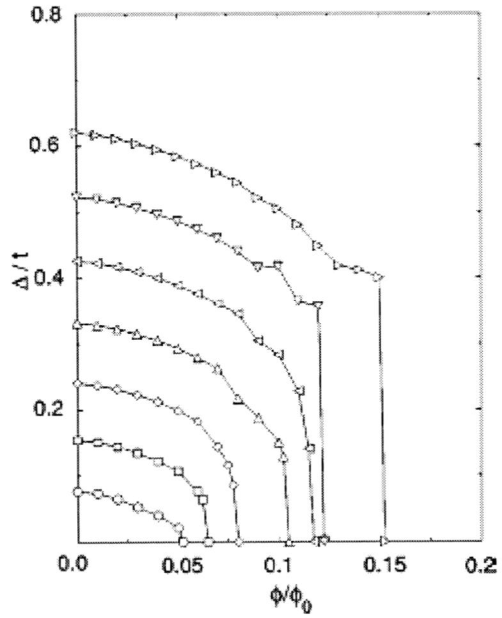

Figure 19. The DDW gap magnitude as a function of the applied flux ϕ/ϕ_0 for various values of the coupling strength V/t :0.5,0.75, 1.0,1.25,1.5,1.75,2.0 from bottom to the top.

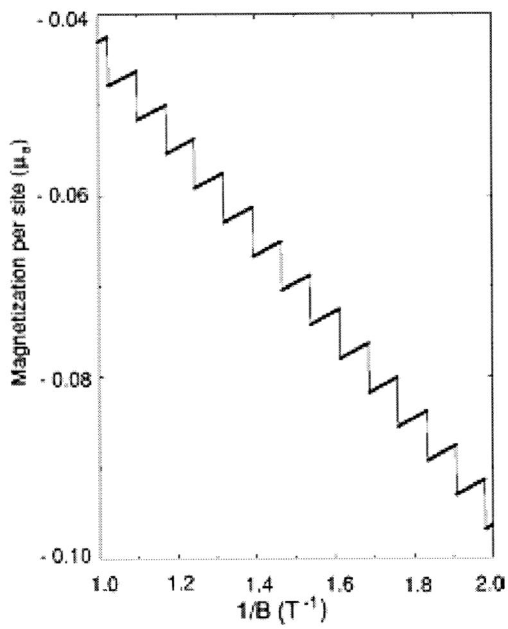

Figure 20. The de Haas-van Alphen oscillations of the magnetizations in the DDW state.

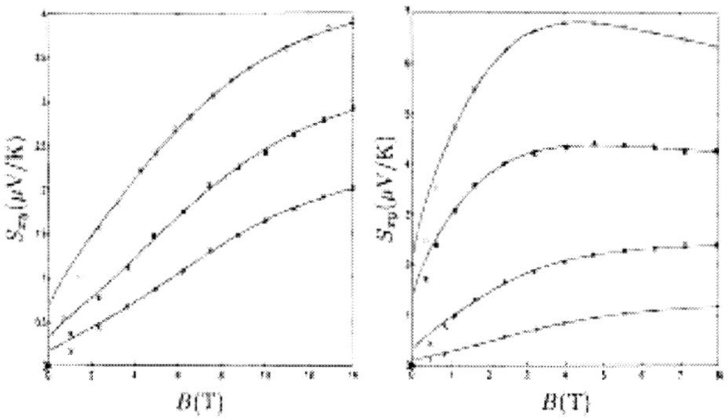

Figure 21. Nernst signal is shown for $La_{2-x}Sr_xCuO_4$ with $x = 0.12$ (left panel) and $Bi_2Sr_{2-y}La_yCuO_6$ with $y = 0.5$ (right panel). The temperatures are 30K, 35K and 40K from the top in the left panel while in the right they are 25K, 35K and 40K

where S is the entropy, B is the magnetic field and ρ is the conductivity. The effect of the magnetic field can be included via the Peierls-Onsager substitution. The details can be found in the works of ref. [71, 75, 76].The Nernst signal in underdoped $La_{2-x}Sr_xCuO_4$ for $x = 0.12$ and $Bi_2Sr_{2-y}La_yCuO_6$ for $y = 0.5$ are plotted and compared with their respective experimental values [77] in fig (21). We observe an excellent agreement between the experimental measurements and the theoretical preditions. The observation of the Nernst signal in the DDW state strengthens further the candidature of DDW state for the pseudogap.

In the above two cases we consider the effect of magnetic field only on the DDW state. In this section, we would like to discuss the role of magnetic field on the three types of states, the DDW, DDW+DSC and DSC. This is important for the understanding of the superconducting phase and the normal phase in the underdoped region [72]. The applied magnetic field, in this case, considers the coupling of the magnetic field to the quasi-particle spin and modifies the hopping term. As the paramagnetism is an observable phenomenon, it justifies the study.

The Hamiltonian is an extension of the eq. (3) [59] and can be written as

$$\begin{aligned} H &= \sum_{\mathbf{k}\sigma}(\epsilon_{\mathbf{k}} - \mu)c^{\dagger}_{\mathbf{k}\sigma}c_{\mathbf{k}\sigma} + \sum_{\mathbf{k}\sigma} W_{\mathbf{k}}c^{\dagger}_{\mathbf{k}\sigma}c_{\mathbf{k}+\mathbf{Q}\sigma} + h.c + \sum_{\mathbf{k}} \Delta_{\mathbf{k}}c^{\dagger}_{\mathbf{k}\uparrow}c^{\dagger}_{-\mathbf{k}\downarrow} + h.c \\ &+ \mu_B \sum_{i\sigma\sigma'} c^{\dagger}_{i\sigma}(\sigma_{\sigma\sigma'}.\mathbf{H})c_{i\sigma'} \end{aligned} \tag{20}$$

where $\sigma_{\sigma\sigma'}$ are the components of the Pauli matrices: $\sigma = (\sigma_1, \sigma_2, \sigma_3)$.

Let us consider the magnetic field to be perpendicular to the c-axis. The inclusion of the magnetic field along the planes will increase the computational difficulties due to increase of the dimension of the matrix. The two mean field equations obtained from the stationary conditions are

$$\Delta_0 = \frac{V_{DSC}}{2}\sum_{\mathbf{k}}\sum_{s_2=\pm}\sum_{s_3=\pm} s_3 \frac{\Delta_0 \eta^2_{\mathbf{k}}}{g_{s_2}(\mathbf{k})} \frac{1}{1+e^{E_\nu(\mathbf{k})/T}} \tag{21}$$

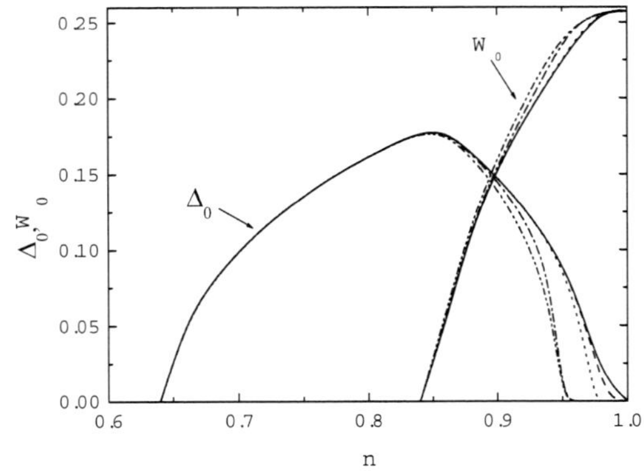

Figure 22. The order parameters vs filling phase diagram for $\beta = 1.0$, $\gamma = 0.0$ and various magnetic field $\mu_B H$=0.0 (solid line), 0.01 (dashed line), 0.05 (dotted line), 0.1 (dashed-dotted line) and 0.12 (dashed-double-dotted line).

and

$$W_0 = \frac{V_{DDW}}{2} \sum_{\mathbf{k}} \sum_{s_2=\pm} \sum_{s_3=\pm} \frac{s_3 W_0 \phi_{\mathbf{k}}^2}{g_{s_2}(\mathbf{k})} \left\{ 1 + s_2 \frac{1}{\sqrt{\epsilon_{\mathbf{k}}^2 + W_{\mathbf{k}}^2}} \right\} \frac{1}{1 + e^{E_\nu(\mathbf{k})/T}} \tag{22}$$

where

$$E_\nu(\mathbf{k}) = \mu_B H \pm \left[\epsilon_{\mathbf{k}}^2 + \mu^2 + \Delta_{\mathbf{k}}^2 + W_{\mathbf{k}}^2 \pm 2\sqrt{\epsilon_{\mathbf{k}}^2 \mu^2 + \mu^2 W_{\mathbf{k}}^2} \right]^{\frac{1}{2}} \tag{23}$$

In short notation we can write $E_\nu(\mathbf{k}) = \mu_B H + s_3 g_{s_2}(\mathbf{k})$ where s_2 and s_3 can take values of ± 1. In fig. (22) the order parameters Δ_0 and W_0 are plotted against filling n. The values of V_{DDW} and V_{DSC} are chosen in a manner so as to obtain a phase diagram which has the DDW, DDW+DSC and the DSC phase in the appropriate regions of doping. It is important to observe the above three phases in order to obtain a phase diagram consistent with the experiments. In general, varying these constants by 10-20% does not change the phase diagram qualitatively. However, larger variations may lead to only the DDW phase or the DSC phase. The sets of lines which represent the DSC gap and the DDW are indicated in the graphs by Δ_0 and W_0 respectively. In the region of doping close to zero, a dominant d-density-wave state is observed. However, away from $\delta = 0.0$ the superconducting order parameter appears in accordance with the experimental predictions. A region of coexistence of the above two types of order parameters is observed in the region $0 \leq \delta \leq 0.16$ which depends on the value of H . The DDW state extends down to half filling as no measure has been taken to observe an insulating phase at $\delta = 0$. The relative values of the superconducting gap or the density wave gap can be altered changing the values of V_{DDW} and V_{DSC}. Moreover, for the same value of doping, the increase in the magnetic field increases DDW order parameter whereas a supression of the DSC gap takes place. Moreover, Δ_0 to T_c ratio increases with the magnetic field despite a decrease in T_c. At $H = 0.0$, Δ_0 to T_c ratio is found to be 1.21. However, when $\mu_B H = 0.1$, the ratio is observed to be 1.43 and for $\mu_B H = 0.12$, the value is 1.51.

Fig. (23) displays the dependence of the order parameters on δ for the orthorhombic lattice; $\beta = 0.95$. Orthorhombicity as low as 2% has been observed in $La_{2-x}Sr_xCuO_4$ [78]

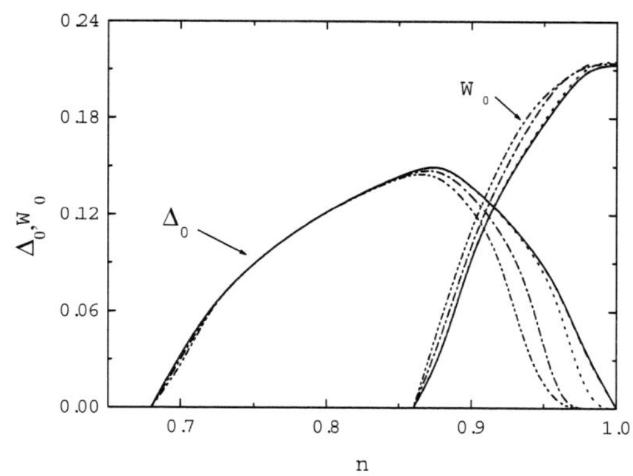

Figure 23. The order parameters vs filling phase diagram for $\beta = 0.95$, $\gamma = 0.0$ and various magnetic field $\mu_B H$=0.0 (solid line), 0.01 (dashed line), 0.05 (dotted line), 0.1 (dashed-dotted line) and 0.12 (dashed-double-dotted line).

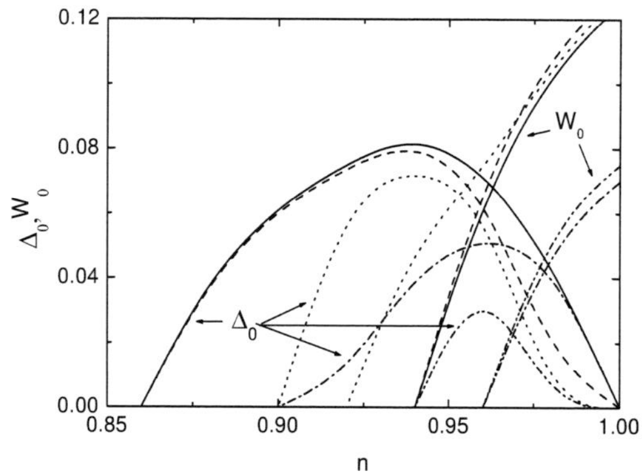

Figure 24. The order parameter vs filling phase diagram for $\beta = 1.0$, $\gamma = 0.1(\mu_B H$=0.0 (solid line), 0.05 (dashed line) and 0.1 (dotted line)) and 0.2(0.0 (dashed-dotted line) and 0.05 (dashed-double-dotted line))).

and can have even higher values. So let us choose a weak orthorhombicity of 5%. The other parameters are same as those in fig (22). The phase diagram also shows a region of co-existence. The region over which both the order parameters exist decreases with the inclusion of β and the effect of the magnetic field becomes more pronounced.

In fig. 24 the order parameters are plotted against filling n in the presence of the next-nearest hopping parameter. A next-nearest neighbor interaction is introduced as done in various theoretical studies of the cuprates [60]. For $\gamma = 0.1$ we observe a sharper fall of W_0. Moreover, the superconducting phase ceases to exist at a higher value of n. Hence, a much narrower coexistence of DDW and the DSC phases occurs when compared to the $\gamma = 0.0$ case. Thus with the inclusion of γ both the DDW and the DSC order parameters decreases and a much smaller value of the magnetic field destroys superconductivity.

In fig.(25) the critical field vs critical temperature T_c is plotted and in fig. (26) the field vs n phase diagram is exhibited. In fig.(25) the critical magnetic field is observed to have a robust power law dependence on the critical temperature: $H_c \sim T_c^2$. Moreover, the external

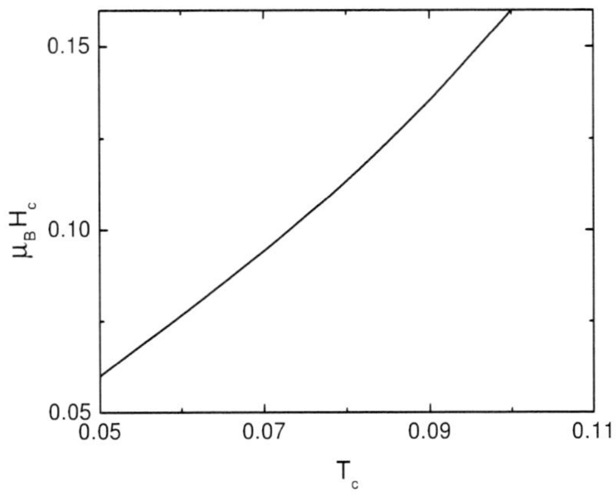

Figure 25. Critical field vs T_c.

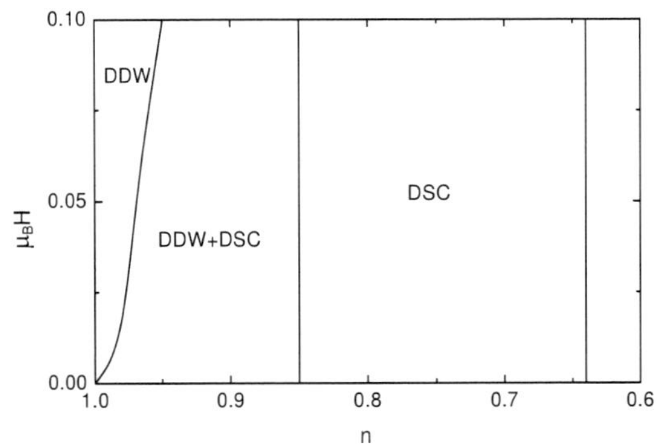

Figure 26. Critical field vs n.

magnetic field (H_c) to T_c ratio is observed to be very similar to the paramagnetic limit but is not a constant as the Δ_0 to T_c ratio increases with the magnetic field. In fig.(26) the effect of the magnetic field on n, as we go away from half filling to underdoped region, is clearly observed for $\beta = 1.0$ and $\gamma = 0.0$. As the magnetic field increases the value of n for the DDW $\rightarrow$ DDW+DSC transition decreases, while that for the DDW+DSC $\rightarrow$ DSC phase transition remains unaltered. The coexistence region of DDW and DSC decreases with H and the decrease occurs at a much faster rate as H increases. This demonstrates the difference in the behavior of the two order parameters with H. Moreover, after a certain value of H we expect only the pure DDW state to prevail. The DDW state is expected to vanish at a much higher value of H as observed in ref [70].

The temperature dependence of the superconducting specific heat $C(T)$ can be observed in fig. (27). With the increase in the field a decrease in the jump of the specific heat is observed. But the $C_s(T)/C_n(T_c)$ ratio increases due to the increase in the Δ_0/T_c ratio with the field. A similar feature showing the decrease in jump can be seen in ref [79].

Hence, from the above study we see that the magnetic field increases the DDW order parameter, it suppresses the superconducting gap and thus the critical temperature. The Δ_0/T_c ratio increases with the field. The coexistence region of DDW and DSC decreases with the magnetic field and also with γ and β. So it is seen that the effect of the field on

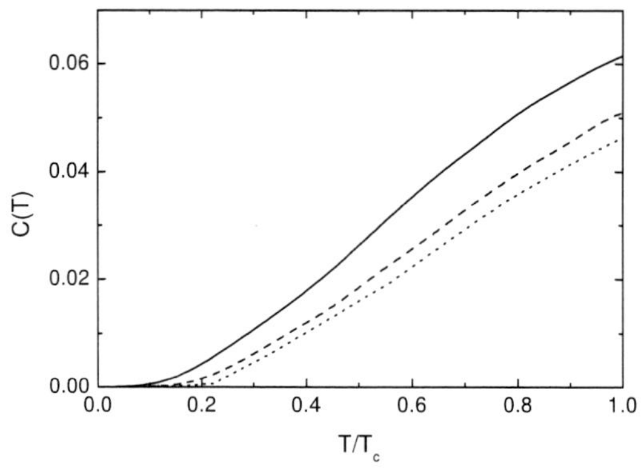

Figure 27. Specific heat vs T/T_c for different magnetic fields.$\mu_B H$=0.0 (solid line), 0.1 (dashed line) and 0.12 (dotted line).

DDW is very much different when compared to the DSC and also the effect is weaker in the case of DDW which is due to the difference of nature of the two condensates. Hence, the effect of the magnetic field on DDW phase and the mixed DSC+DDW phase are in accord with the experimental predictions.

In the next part of the discussion we consider the penetration depths in the underdoped region of the superconductors. First, let us calculate the penetration depth λ along the a and b directions in the presence of the pseudogap and superconducting phases for a simple model that incorporates two layers - a CuO_2 plane and a CuO chain per unit cell [94]. It is widely believed that the source of the pairing interaction which is responsible for the superconducting transition lies in the CuO_2 planes and hence could be explained by the two-dimensional model. The only active pieces of the crystal are the CuO_2 planes and the remaining ions act as charge reservoirs. But in some materials, such as the $YBa_2Cu_3O_7$ (Y-123) and $YBa_2Cu_4O_8$ (Y-124), there is clear evidence of the fact that CuO_2 planes are not the only active portions. In these materials we have the quasi one-dimensional CuO chain structures. Experiments on the dc resistivity, [80, 81] the polarized reflectance, [82, 83] and penetration depth [83–85] have found large anisotropies between the a and the b directions which suggest that currents also flow along the chains in both the normal and superconducting states. But little is known about the coupling of the planes and the chains and how the planar superconductivity affects the physics in the chains. Recently, the proximity model [86] has been demonstrated to explain consistently the scanning tunneling microscopy, angle-resolved photoemission, and infrared-spectroscopy experiments.The CuO chains become superconducting due to the proximity to the planes below the critical temperature. The model is different from our earlier Hamiltonian given in eq. (3) due to the consideration of the proximity model which is important for some of the cuprates. The pairing interaction is considered to be solely localized to the CuO_2 planes. The k dependence of the plain-chain coupling is neglected, since this term is important in a very small k space region. Hence, omitting the k_z dependence initially, we can write down the Hamiltonian as

$$H = \sum_{\mathbf{k}\sigma}(\epsilon_{\lambda\mathbf{k}} - \mu)c^\dagger_{\lambda\mathbf{k}\sigma}c_{\lambda\mathbf{k}\sigma} + \sum_{\mathbf{k}\sigma} t(k)c^\dagger_{1\mathbf{k}\sigma}c_{2\mathbf{k}\sigma} + \sum_{\mathbf{k}\sigma} W_{\mathbf{k}}c^\dagger_{1\mathbf{k}\sigma}c_{1\mathbf{k}+\mathbf{Q}\sigma}$$

$$+ \quad \sum_{\mathbf{k}} \Delta_{\mathbf{k}} c^{\dagger}_{1\mathbf{k}\uparrow} c^{\dagger}_{1\mathbf{k}\downarrow} + h.c \tag{24}$$

where $c_{1\mathbf{k}\sigma}$ is the fermion creation operator for wave vector k and spin σ in the plane and $c_{2\mathbf{k}\sigma}$ is the same for the chain band. The energy dispersion relations are given by $\epsilon_{1k} = -2t_1[\cos(k_x a) + \cos(k_y b)]$ and $\epsilon_{2k} = -2t_2 cos(k_y b)$ where a and b are the lattice constants in the planes. μ_λ is the chemical potential for the plane and the chain bands. $t(k)$ is the coupling between the planes and the chains. λ corresponds to the chain and/or plane bands.

$$\lambda_{ij}^{-2} = \frac{8\pi e^2}{\hbar^2 c^2} \frac{1}{\Omega} \sum_{\mathbf{k},\alpha\beta} \hat{\gamma}^{(i)}_{\mathbf{k}\alpha\beta} \hat{\gamma}^{(j)}_{\mathbf{k}\beta\alpha} \left(\hat{G}_{\mathbf{k}\alpha\beta} \mid_{\Delta=0} - \hat{G}_{\mathbf{k}\alpha\beta} \right) \tag{25}$$

where

$$\hat{G}_{\mathbf{k}\alpha\beta} = \frac{df(E_{\mathbf{k},\alpha})}{dE_{\mathbf{k},\alpha}} \delta_{\alpha\beta} + \frac{f(E_{\mathbf{k},\alpha}) - f(E_{\mathbf{k},\beta})}{E_{\mathbf{k},\alpha} - E_{\mathbf{k},\beta}} (1 - \delta_{\alpha\beta}) \tag{26}$$

The matrices $\gamma_\mu(\mathbf{k}, \mathbf{k})$ are determined by the same approach as used in the work of [89] where the matrices $\hat{\gamma}_\mu(\mathbf{k}, \mathbf{k})$ are given by

$$\hat{\gamma}_\mu(\mathbf{k}, \mathbf{k}) = U^\dagger(\mathbf{k}) \gamma_\mu(\mathbf{k}, \mathbf{k}) U(\mathbf{k}) \tag{27}$$

$E_{\mathbf{k}\alpha}$ are the eigenvalues, $f(x)$ is the Fermi function, $\delta_{\alpha\beta}$ is a kronecker delta and $U_{\mathbf{k}\alpha\beta}$ is the unitary matrix which diagonalizes the Hamiltonian. $E_{\mathbf{k}\alpha}$ and $U_{\mathbf{k}\alpha\beta}$ are determined by numerical techniques.

In the following figures the question of anisotropy and the candidature of the DDW phase within the scenario of the proximity model will be dealt with. The parameters of the model are considered so as to obtain $\lambda_a^2(0)/\lambda_b^2(0) = 1.8$ as observed experimentally in [83].

In fig. (28) the dependencies of T_c and T^* on t are exhibited. The critical temperature is observed to decrease rapidly with increasing t and approaches zero at $t = 0.15$. But the T^* does not seem to be affected by t at such a rapid rate. In the absence of the pseudogap the T_c is much higher; and hence a much higher value of the coupling between the chain and the plane bands can be applied before the disappearance of the superconductivity. The competition between the DDW and the DSC decreases T_c in the presence of the pseudogap as expected. Next, the penetration depths on the plane along the a and b directions are considered to study the anisotropy due to the consideration of the chain bands in the DDW model.

In fig. (29) λ_a and λ_b are demonstrated in the mixed phase, i.e., where the DDW and the DSC coexist in the underdoped region for $\mu_1 = -0.39$ with $T_c = 0.021$, $T^* = 0.16$, and the pure superconducting wave after the DDW has already disappeared in the overdoped regime for $\mu_1 = -0.6$ and $T_c = 0.12$. The above plots are done for $t = 0.0$. We observe an anisotropy in the penetration depth in the mixed phase and also in the pure superconducting phase even in the absence of the interaction between the two layers. The change in the nature of the graph from underdoped to the overdoped case is similar to that of the results of Ref. [87] where they make measurements of the planar penetration depth for various doping which should qualitatively depict the a-axis penetration depth of our model. The anisotropy in the chains and its values are a clear indication of the dissipation in the chains. So it is seen that the DDW model incorporating the chains and the plane can explain the

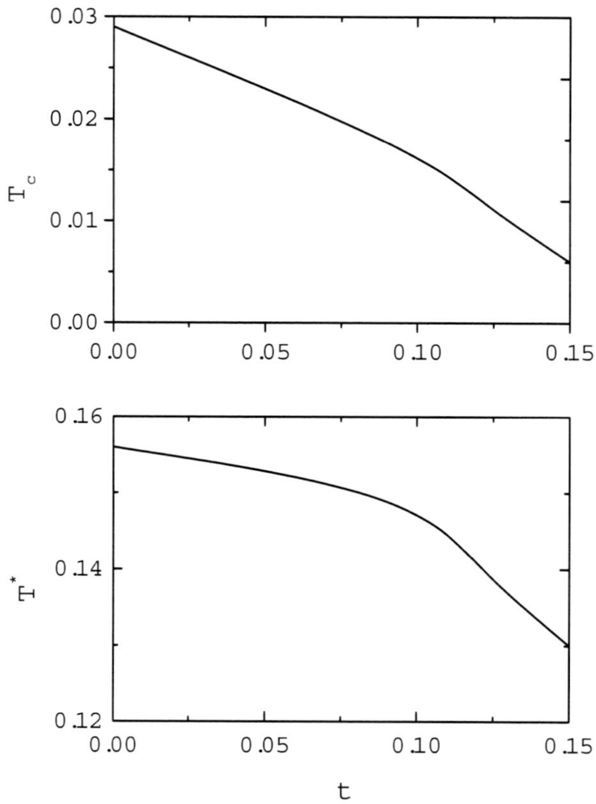

Figure 28. (a)T_c vs t (b) T^* vs t

penetration depth of these superconductors. The anisotropy observed in the results also throws light on the understanding of the universal correlation between T_c and the superfluid density as suggested by Uemura et al. [88] in materials with copper chains.

Next, the effect of the interaction between the layers can be observed in fig. (30). In fig. 30(a) λ_a and λ_b are plotted in the pure DSC phase for $t = 0.10$ with $T_c = 0.12$ and in fig. 30(b) the penetration depths for the mixed DDW+DSC phase are plotted in the underdoped regime. In the mixed phase we have $T_c = 0.017$ and $T^* = 0.15$. The ratio of $\lambda_a^2(0)/\lambda_b^2(0) = 1.8$ at $T = 0.0$ which is close to the value observed in experimental measurements for the pure DSC state [83]. The temperature dependence has a distinct character when compared to the mixed state and is very similar to the calculation in the pure state in Ref. [89]. The introduction of the coupling decreases the value of λ_a and λ_b irrespective of the doping. In fig. 30(c) and fig. 30(d) for $t = 0.12$ we observe similar traits as observed in fig. 30(a) and fig. 30(b). Hence, it can be seen that this simple model can provide some insight on the penetration depth results as we go from the underdoped to the overdoped region of the phase diagram along the a axis and also along the chains of the superconductor to understand the effect of the DDW and the chains on the values of λ. Both the pseudogap phase and t define the nature of temperature dependence. In the range of temperatures close to T_c the critical phase fluctuations dominate the low-frequency electromagnetic re-

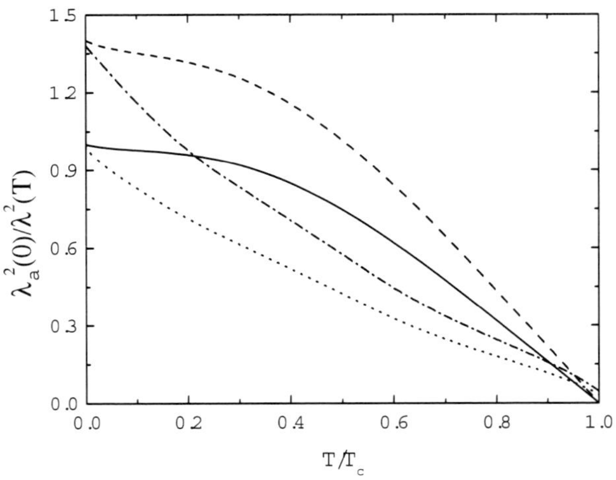

Figure 29. $\lambda_a^2(0)/\lambda^2(T)$vs T/Tc for $t = 0.0$.The solid line (dashed line) indicates the λ_a (λ_b) for the DDW+DSC phase whereas the dotted (dashed-dotted) line indicates the λ_a (λ_b) for the pure DSC phase.

sponse and a low temperature regime the superfluid density has an anomalous temperature dependence as a result of the phase fluctuations to first order in T. However, away from the critical points we expect the results to be correct as already observed in the calculation of superfluid density [63] and its agreement with experimental measures [90]

In the above study the penetration depths in a and b directions are calculated without the inclusion of the momentum depedence of $t(k)$. However, the coherent tunneling in the c-direction with a phenomenological momentum-dependent factor to account for the doping dependences in the c-axis penteration depth has been considered very recently [91]. In the following part the main aim is to emphasize both the doping and the temperature dependence of the a, b and c-axes penetration depth [92]. This model combines the tunneling in the c-direction with a momentum dependent factor to account for the the proximity effect. The chains may play a key role in the qualitative behavior of penetration depth. Anisotropy in the experimental prediction of $\lambda(T)$ has been an additional motivation of this study. On a more general basis the interest is to understand the effect of the momentum dependent chain-plain coupling in the underdoped regime of these d-wave superconductors. $t(\mathbf{k}) = -t\cos(k_z d/2)$ is the coupling between the chains and the planes, where $d/2$ is the distance between the chains and the planes. $t(\mathbf{k})$ not only determines the gap in the chains but also determines the ratio of $\lambda_c(T = 0)/\lambda_{ab}(T = 0)$.

In fig. (31) the temperature dependencies of the penetration depths along the a, b and c directions for various values of t are exhibited. It can be seen that $t = 0.05$, 0.1, 0.12 and 0.15 in fig 31(a), (b), (c) and (d) respectively. We observe as in ref [93] that both $\lambda_a(T)$ and $\lambda_b(T)$ are linear upto a certain value of T/T_c. The value of T/T_c upto which the linearity relation holds depends on the value of t. For example at $t = 0.05$, λ_a and λ_b are linear upto $T/T_c = 0.18$ whereas for $t = 0.15$ we obtain $T/T_c = 0.44$ which can be observed in ref. [93]. A change in the qualitative behavior of the temperature dependence of $\lambda_a(T)$ and $\lambda_b(T)$ is observed from that of the previous work [94]. However, $\lambda_a^2(0)/\lambda_b^2(0) \sim 1.8$

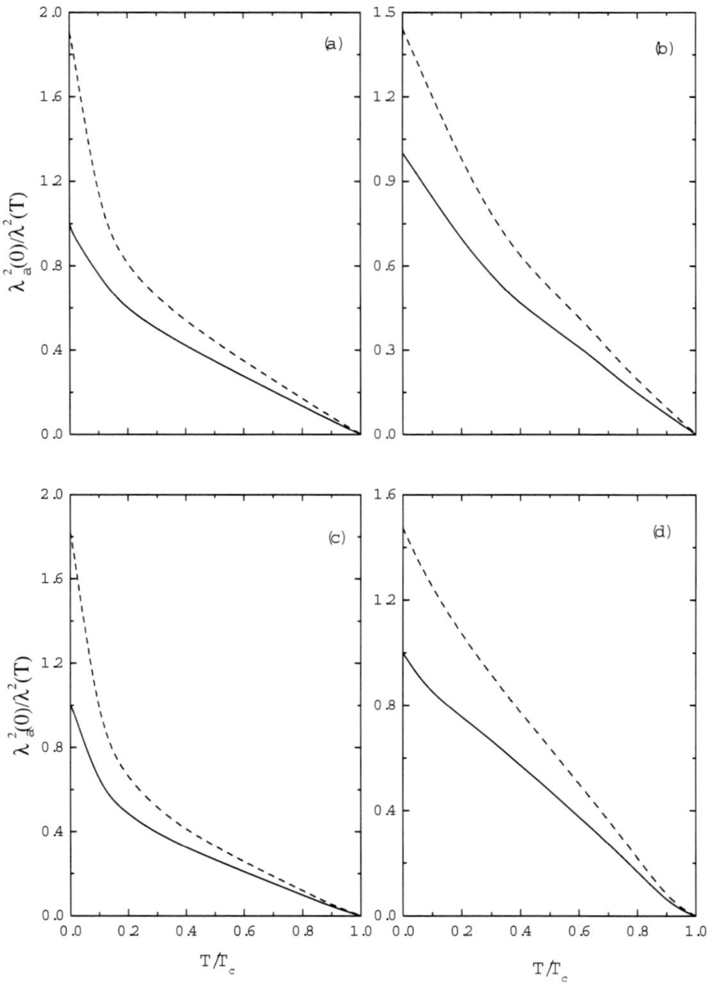

Figure 30. $\lambda_a^2(0)/\lambda^2(T)$vs T/Tc for $t = 0.1$. (a) Pure DSC state and (b) mixed DDW+DSC phase. $\lambda_a^2(0)/\lambda^2(T)$ vs T/Tc for t=0.12, (c) pure DSC state, and (d) mixed DDW+DSC phase. The solid line (dashed line) indicates the λ_a (λ_b).

for $t_0 = 0.0$ as in ref. [94]. For $t = 0.05$, it can be seen that $\lambda_a^2(0)/\lambda_b^2(0) \sim 1.7$ and it decreases to 1.5 when $t = 0.1$. The critical temperatures are 61.2K for $t = 0.05$, 46.8K for $t = 0.1$, 32.4K for $t = 0.12$ and 28.8K for $t = 0.15$. Comparing the present values of T_c and $\lambda_a^2(0)/\lambda_b^2(0)$ with those obtained in the previous work [94] it can be observed that the effect of the inclusion of the momentum dependence in the chain-plane coupling is negligible. The temperature dependence of λ_a, λ_b and λ_c are in accord with the experimental results [93, 95]. The penetration depth $\lambda_a(T)$ and $\lambda_b(T)$ differ qualitatively from the results in ref. [94].

In fig. (32) the temperature dependencies of λ_a, λ_b and λ_c for $t_0 = 0.05$ for various values of T_c are shown. The values of T_c are 18K, 36K and 61.2K which depends on the doping and can be obtained by varying the parameter μ_1. The experimental results [95] for $T_c = 18.2$K are plotted to compare with the theoretical predictions and we observe a very

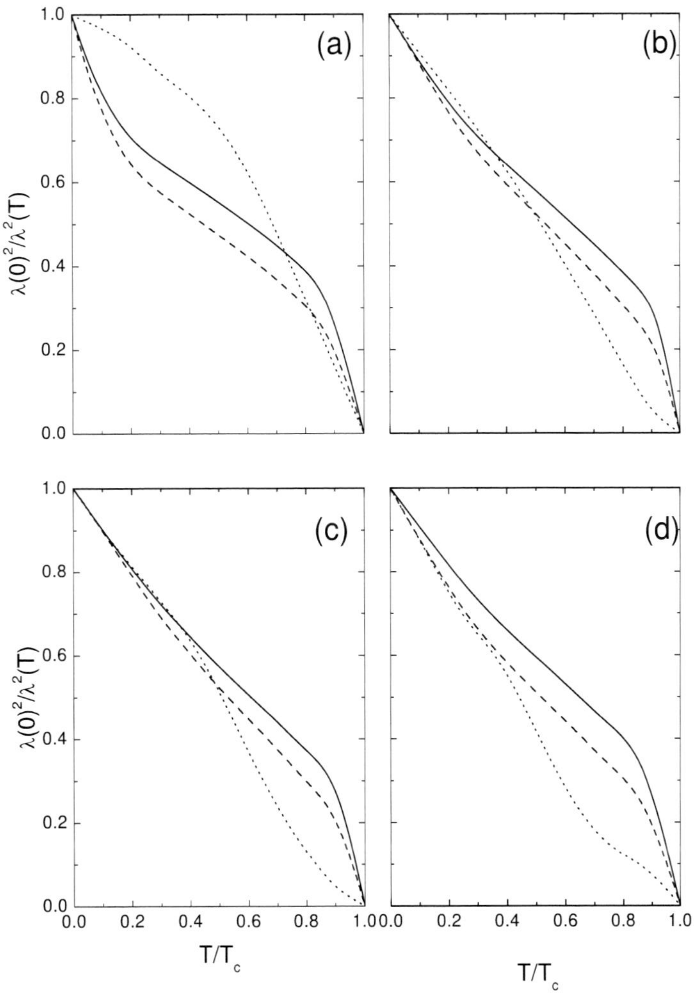

Figure 31. $\lambda^{-2}(T)/\lambda^{-2}(0)$ vs T/T_c along the a (solid line), b (dashed line) and c (dotted line) directions for (a) $T_c = 61.2$K (b)$T_c = 46.8$K (c) $T_c = 32.4$K and (d) $T_c = 28.8$K

good agreement. The plots for $\lambda_c^{-2}(T)$ when plotted with T satisfies the same qualtitative behavior as observed in [95] for the three values of T_c. Comparing the T_c of the three sets and the $\lambda_c^{-2}(0)$ a similar relation between the experimental measurements and our work is observed. Close to T_c a deviation from the experimental results is observed where critical phase fluctuations dominate the low frequency electromagnetic response which has not been considered in the model. In the present work it can be observed that when $T_c^1/T_c^2 \sim 2, 1.7$ and 3.4 the ratios are $\lambda_{c1}^{-2}(0)/\lambda_{c2}^{-2}(0) \sim 3.4, 2.5$ and 8.28 respectively. Comparing it with the experimental values for nearly same ratios of T_c's similar values are obtained. From ref. [95] the ratios are found to be $\lambda_{c1}^{-2}(0)/\lambda_{c2}^{-2}(0) \sim 3.0, 1.7$ and 8.0 when $T_c^1/T_c^2 \sim 2.2, 1.23$ and 2.7 respectively. A similar trend is observed in the values as the ratio of the critical temperature decreases or increases. So, not only the temperature dependencies but also the absolute values follow the experimental predictions.

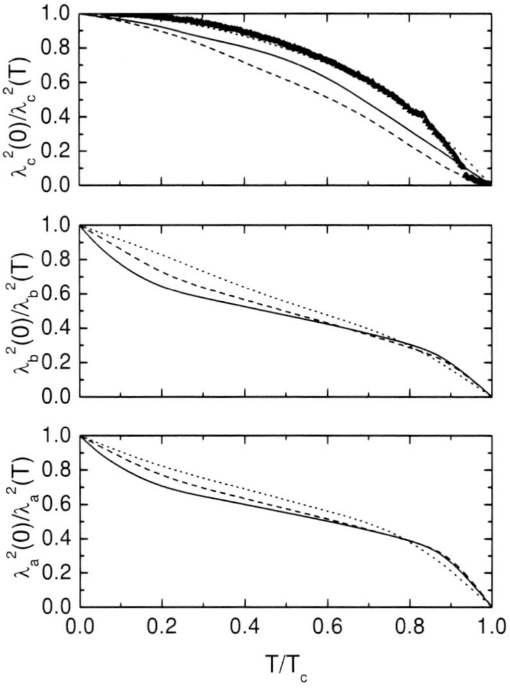

Figure 32. $\lambda^{-2}(T)/\lambda^{-2}(0)$ vs T/T_c along the (a) a, (b) b and (c) c directions for $T_c = 61.2$K (solid line) $T_c = 36.0$K (dashed line) and $T_c = 18.0$K (dotted line). The symbols in (c) denotes the experimental results from ref. [95]

A qualitative difference is observed with the earlier results [94] in the temperature dependence of the λ_a and λ_b. The c-axis penetration depth λ_c is calculated and observed to have a very good agreement with the experimental result. So the simple model with the chain-plane coupling is able to describe qualitatively the temperature dependence of the penetration depth along the three directions in the underdoped region where we have the DDW order and the DSC order. The results demonstrate that the experimental results can be used to distinguish between the underlying phases – a DDW state, DSC state or a co-existing DDW and DSC state.

Let us now conclude the discussion on the properties of the superconductors in the pseudogap phase with the study of the Hall coefficiecient R_H [96]. The preliminary experimental measurements [97] indicate a sharp signature in $n_{Hall} = 1/R_H$ at a carrier concentration of 0.15 which is completely in conflict with the conventional metals where we can write $n_{Hall} = ne/B$ where n is the carrier density, e is the electron charge and B is the applied magnetic field. The mean field Hamiltonian for the DDW case given by

$$H = \sum_{\mathbf{k}\sigma}(\epsilon_{\mathbf{k}} - \mu)c^{\dagger}_{\mathbf{k}\sigma}c_{\mathbf{k}\sigma} + \sum_{\mathbf{k}\sigma} W_{\mathbf{k}}c^{\dagger}_{\mathbf{k}\sigma}c_{\mathbf{k}+\mathbf{Q}\sigma} + h.c \tag{28}$$

is considered and the Kubo formulae is applied to calculate n_{Hall}. The results shown in fig. (33) correspond to $t = 0.3eV$, $\gamma t/t = 0.3$, $\mu = -t$ and carrier concentration $x_c = 0.2$ and observed to be in accord with the experimental result [97]. Recently the infrared hall measurements were dealt with using the mean field Hamiltonian for the DDW case [98].

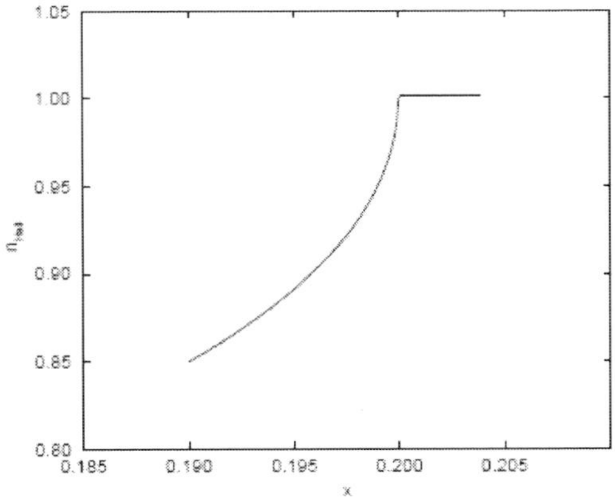

Figure 33. Hall number vs doping

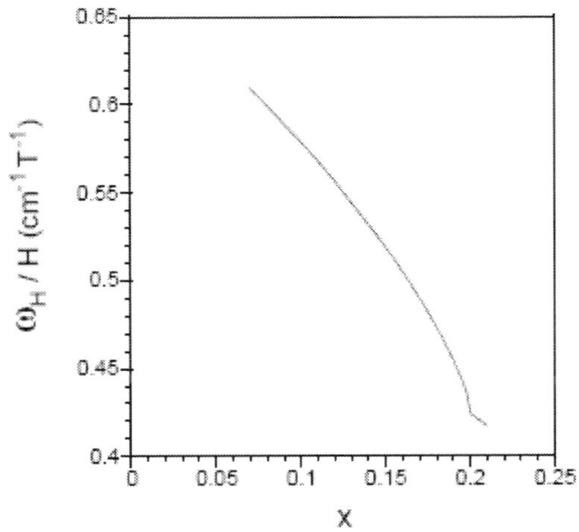

Figure 34. Hall frequency vs x.

The hall frequency ω_H is plotted as a function of the hole doping x in fig. (34) and the results are found to be in agreement with the experimental measurements [99].

4. Conclusion

In this article we have given a discussion on the role of the DDW order in the phase diagram of the cuprates. Despite the ongoing controversy regarding the origin of the pseudogap in the cuprates, the manuscript poses DDW order as a strong candidate. The phase diagram of the cuprates could be very well explained which demonstrates clearly the competition of the two phases. Moreover, the study of the specific heat, spectral function, density of states, penetration depths have been found to be in accord with the experimental measurements. However, the experiments in the underdoped region continue to be not only few in number but also depict a very anomalous view of this region. Hence, the origin of the pseudogap is still far from reaching a general consensus. Further detailed theoretical and experimental studies could help us decipher the mystery of this phase. The objective of this article is to demonstrate the success of the DDW phase in explaining the pseudogap. The studies till date have proven to be of substantial importance as can be seen from the above discussion of the phase diagrams and various properties in the underdoped phase of the cuprates.

The author acknowledges Dr. P. Chaudhuri for helpful discussions.

References

[1] Bednorz, J. G.; Muller, K. A.; *Z. Phys. B: Condens. Matter* (1986) 64, 189.

[2] Wu, M. K.; Ashburn, J. Torng, C.; Hor, P.; Meng, R.; Gao, L.; Huang, Z.; Wang, Y.; Chu, C.; *Phys.Rev. Lett.* (1987) 58 9078.

[3] Bardeen, J.; Cooper, L. N.; Shrieffer, J. R.; *Phys. Rev* (1957) 108, 1175.

[4] Gough, C. E,; *Nature* (1987) 326, 855.

[5] Adhikari, S. K.; Ghosh, A.;*J. Phys.: Cond. Matter* (1998) 10 135.

[6] Tsuei C. C.; Kirtley J. R.; *Rev. Mod. Phys.* (2000) 72 969.

[7] Hardy W. *et al.*; *Phys. Rev. Lett.* (1993) 70, 3999.

[8] Moler K. A.*et al.*; *Phys. Rev. Lett.* (1994) 73, 2744.

[9] Ruckenstein, A. E.; Hirschfeld, P. J.; Apel, J.; *Phys. Rev. B* (1987) 36 857.

[10] Kotliar, G.; *Phys. Rev. B* (1988) 37 3664.

[11] Sun, A. G.; Gajewski, D. A.; Maple, M. B.; Dynes, R. C.; *Phys.Rev. Lett.* (1994) 72 2267.

[12] Chaudhari, P.; Lin, S. Y.; *Phys. Rev. Lett.* (1994) 72 1084.

[13] Xu, J. H.; Shen, J. L.; Miller Jr., J. H.; Ting, C. S.; *Phys. Rev. Lett.* (1994) 73 2492.

[14] Li, Q. P.; Koltenbah, B. E. C.; Joynt, R.; *Phys. Rev. B* (1993) 48 437.

[15] Ghosh, A.; Adhikari, S. K.; *Physica C* (1999) 322 37.

[16] Krishana, K.; Ong, N. P.; Li, Q.; Gu, G. D.; Koshizuka, N.; *Science* (1997) 277 83.

[17] Laughlin, R. B.; *Phys. Rev. Lett.* (1998) 80 5188.

[18] Ghosh, H.; *Europhysics Lett.* (in press).

[19] Franz, M.; Tešanović, Z.; Tešanović, *Phys. Rev. Lett.* (1998) 80 4763.

[20] Ghosh A.; Adhikari S. K.; *Phys. Rev. B* (1999) 60 10401.

[21] Timusk, T.; and Statt, B.; *Rep. Prog. Phys.* (1999) 62, 61.

[22] Tallon, J. L.; Loram, J. W., *Physica C* (2001) 349, 53.

[23] Warren Jr, W. W.; Walstedt, R. E.; Brennert, J. F.; Cava, R. J.; Tycko, R.; Bell, R. F.; Dabbagh, G.; *Phys. Rev. Lett.* (1989) 62 1193.

[24] Bankay, M.; Mali, M.; Roos, J.; Brinkmann, D.; *Phys. Rev. B* (1994) 50 6416.

[25] Stern, R.; Mali, M.; Mangelschots, I.; Roos, J.; Brinkmann, D.; Genoud, J-Y.; Graf, T.; Muller, J.; *Phys. Rev. B* (1994) 50 (1994) 426.

[26] Walstedt, R. E.; Bell, R. F.; Mitzi, D. B.; *Phys. Rev. B* (1991) 44 7760.

[27] Julien, M. H.; Caretta, P.; Horvatic, M.; Berthier, M.; Berthier, C.; Segransan, Y.; Carington, P.; Colson, D.; *Phys. Rev. Lett.* (1996) 76 4238.

[28] Alloul, H.; et al. *Phys. Rev. Lett* (1990) 67 3140.

[29] Williams, G. V. M.; et al.*Phys. Rev. B* (1995) 51 16503.

[30] Norman M. et. al. *Nature* (1998) 392, 157.

[31] Rennner, Ch.; Revaz, B.; Genoud, J-Y.; Kadowaki, K.; Fischer, O.; *Phys. Rev. Lett.* (1998) 80 149.

[32] Loram, J. W.; Mirza, K. A.; Cooper, J. R.; Tallon, J. L.; *Physica C* (1997) 282-287, 1405.

[33] Loram, J. W.; Mirza, K. A.; Cooper, J. R.; Athanassopoulou, N.; Liang, W. Y. *Proc 10th HTS Anniversary workshop on Physics, Materials and Applications* ed. B. Batlogg et al, World Scientific.

[34] Loram, J. W.; et al *Physica C* (2000) 341-348, 831.

[35] Bucher, B.; Steiner, P.; karpinski, J.; Kaldi, E.; Watcher, P. *Phys. Rev. Lett.* (1993), 70, 2012.

[36] Takagi, H.; Batlogg, B.; Kao, H. L.; Kwo, J.; Cava, R. J.;Krajewski, J. J.; Peck, W. F. Jr *Phys. Rev. Lett.* (1992), 69, 2975.

[37] Ito, T.; Takenaka, K.; Uchida, S. *Phys. Rev. Lett.* (1993), 70, 3995.

[38] Hwang, H. Y.; Batlogg, B.;Takagi, H.; Kao, H. L.; Kwo, J.; Cava, R. J.;Krajewski, J. J.; Peck, W. F. Jr.; *Phys. Rev. Lett.* (1994), 72, 2636.

[39] Chien, T. R.; Wang, Z. Z.; Ong, N. P.*Phys. Rev. Lett.* (1991), 67, 2008.

[40] Blumberg, G.; Klein, M. V.; Kadowaki, K.; Kendziora, C. Guptasarma, P.; Hinks, D. *J. Phys. Chem. Solids* (1998), 59, 1932.

[41] Puchkov, A. V.; Fournier, P.; Basov, D. N.; Timusk, T.; Kapitulnik, A.; Kolesnikov, N. N. *Phys. Rev. Lett.* (1996), 77, 3212.

[42] Bourges, P. *The Gap symmetry and Fluctuation in High Temperature Superconductors*, Eds Bok J.; Deutsher G.; Pavuna D.; Wolf S. A.; Plenum Press, New York (1998).

[43] Randeria, M., cond-mat/9710223 (unpublished) and references therein; Sa de Melo, C. A. R.; Randeria, M.; Engelbrecht, J. R.;*Phys. Rev. Lett.* (1993), 71, 3202.

[44] Chen, Q. Stajic, J.; Tan, S.; Levin, K.; Physics Reports (2005), 412, 1.

[45] Anderson, P. W.; *Science* (1987), 235, 1196.

[46] Nagaosa, N.; Lee, P. A.; *Phys. Rev. B* (1992), 45, 966.

[47] Fukuyama H.; Prog. Theo. Phys. Suppl. (1992), 108, 287.

[48] Emery, V. J.; Kivelson, S. A.; Phys. Rev. Lett. (1995) 74, 3253.

[49] Uemura, Y. J.; et. al. *Phys. Rev. Lett.* (1989), 62, 2317.

[50] Pines, D.; *Z. Phys. B* (1997), 103, 129; Pines, D.; *Physica C* (1997), 282-287, 273.

[51] Millis, A. J.; Monien, H.; *Phys. Rev. Lett.* (1993), 70, 2810.

[52] Castellani C.; Di Castro C.; Grilli M. *Z. Phys.* (1997), 130, 137.

[53] Brown, S.; Gruner, G Scientific American (1994), 270, 28.

[54] Taraphder, A.; Krisnamurthy, H. R.; Pandit, R.; Ramakrishnan, T. V.; *Phys. Rev. B* (1995), 52, 1368.

[55] Gabovich, A. M.; Voitenko, A. I.; Ausloos, M; Physics Reports (2002), 367, 583.

[56] Chakravarty, S.; Laughlin, R. B., Morr, D. K.; Nayak, C., *Phys. Rev. B* (2001) 63 9450.

[57] .Dimov, I.; Nayak, C.; cond-mat/0512627.

[58] Nayak, C.; *Phys. Rev. B* (2000), 52, 1368.

[59] Ghosh, A.; *Int. J. Mod. Phys. B*, (2001) 15 4161.

[60] Nayak, C.; Pivovarov, E.; *Phys. Rev. B* (2002), 66, 064508.

[61] Cappelluti, E.; Zeyher, R. *Phys. Rev. B* (1999), 59, 6475.

[62] Loram, J. W.; Mirza, K. A.; Cooper, J. R.; Tallon, J. L.; J. Phys. Chem. Solids (1998) 59, 2091.

[63] Tewari, S.; Lee, H. -Y.; Nayak ,C.; Chakrabarty, S.; *Phys. Rev. B* (2001), 64, 22451.

[64] Morr, D. K. ;*Phys. Rev. Lett* (2004), 93, 089902.

[65] Bena, C.; Chakravarty, S.;Jiangping, H.; Nayak, C.; *Phys. Rev. B* (2004), 69, 134517.

[66] Wang, Q. H.; Lee, D. H.; *Phys. Rev. B* (2003), 67, 020511.

[67] Bena C.; Chakravarty S.; Jiangping H.; Nayak C.; cond-mat/0405468.

[68] Anderson, B. M.; *Phys. Rev. B* (2003), 68, 094518.

[69] Ghosal, A.; Kee, H.; cond-mat/0306415.

[70] Nguyen, H. K.; Chakravarty, S.; *Phys. Rev. B* (2002) 65 180519.

[71] Maki, K.; Dora, B.; Virosztek, A.; Vanyolos, A.; *Curr. App. Phys.* (2004), 4, 693.

[72] Ghosh, A. *Physica B* (2006) 373 240.

[73] Zheng, G-q.; *et. al. Phys. Rev. B* (1999), 60, R9947.

[74] Gorny, K.; *et. al. Phys. Rev. Lett* (1999), 82, 177.

[75] Dora, B.; Maki, K.; Virosztek, A.; *Phys. Pol. B* (2003), 34, 571.

[76] Dora, B.; Maki, K.; Vanyolos, A.;Virosztek, A.; *Phys. Rev. B* (2003) 68 241102.

[77] Wang Y. *et. al. Phys. Rev. B* (2001), 64, 224519.

[78] Matsuda, M.; Fujita, M.;, Yamada, K.; Birgeneau, R. J.; Kastner, M. A.; Hiraka, H.; Endoh, Y.; Wakimoto, S.; Shirane, G.; *Phys. Rev. B* (2000) 62, 9148.

[79] Schilling, A.; Fisher, R. A.; Phillips, N. E.; Welp, N.; Kwok, W. K.; Crabtree, G. W.; *Phys. Rev. Lett.* (2002) 78 4833.

[80] Friedmann, T. A.;Rabin, M. W.;Giapintzakis, J.; Rice, J. P.; Ginsberg, D. M.; *Phys. Rev. B* (1990) 42, 6217.

[81] Gagnon, R.; Lupien C.; Taillefer, L.; *Phys. Rev. B* (1994) 50, 3458.

[82] Koch, B.; Geserich, H. P.; Wolf, Th.; *Solid State Comm.* (1989) 71, 495.

[83] Basov, D. N.; *et. al. Phys. Rev. Lett* (1995) 74, 598.

[84] Tallon, J. L.; Bernhard, C.; Binninger, U.; Hofer, A.; Williams, G. V. M.; Ansaldo, E.J.; Budnick, J. I.; Niedermayer, Ch.; *Phys. Rev. Lett.* (1995) 74, 1008.

[85] Zhang, K.; *et. al. Phys. Rev. Lett.* (1994) 73, 2484.

[86] Morr, D. K.; Balatsky, A. V.; *Phys. Rev. Lett.* (2001), 87, 247002.

[87] Trunin, M. K.; Nefyodov, Y. A.; Shevchun, A. F.; *Phys. Rev. B* (2004) 92, 067006.

[88] Uemura, Y. J.; *Physica C* (1997), 282-287, 194.

[89] Atkinson, W. A.; Carbotte, J. P.; *Phys. Rev. B* (1995) 52, 10601.

[90] Panagopoulos, C.; Tallon, J. L.; Xiang, T.;*Phys. Rev. B* (2001) 59, R6635.

[91] Sheehy, D. E.; Davis, T. P.;Franz, M.; *Phys. Rev. B* (2004) 70, 054510.

[92] Ghosh, A. (communicated, 2006).

[93] Homes, C. C.;Timusk, T.; Bonn, D. A.; Liang, R.; Hardy, W. N.; *Physica C* (1995) 254, 265.

[94] Ghosh A.; *Phys. Rev. B* (2006) 73 012504.

[95] Hossein, A.; Broun, D. M.; Sheehy, D. E.; Davis, T. P.; Franz, M.; Hardy, W. N.; Liang, R.; Bonn, D. A.; *Phys. Rev. Lett.* (2004) 93, 107003.

[96] Chakrabarty, S.; Nayak, C.; Tewari, S.;Yang, X.; *Phys. Rev. Lett.* (2002), 89, 277003.

[97] Balakirev, F. F.; Betts, J. B.; Ono, S.; Murayama, T.; Ando, Y.; Boebinger, G. S.; *Nature* (2003), 424, 912.

[98] Tewari, S.; Chakrabarty, S.; Fjaerestad, J. O.; Nayak, C.; Thompson, R. S.; *Phys. Rev. B* (2004) 70, 014514.

[99] Rigal, L. B.; Schmadel, D. C.; Drew, H. D.; Maiorov, B.; Osquigil, E.; Preston, J. S.; Hughes, R.; Gu, G. D.; *Phys. Rev. Lett.* (2004), 93, 137002.

In: Progress in Superconductivity Research
Editor: Oliver A. Chang, pp. 163-198

ISBN 1-60021-694-3

Chapter 6

MODEL OF SUPERCONDUCTIVITY IN THE SINGULAR FERMI LIQUID

Ryszard Gonczarek[1] ***and Mateusz Krzyzosiak***[1,2*]
[1]Institute of Physics, Wrocław University of Technology
Wybrzeże Wyspiańskiego 27, 50-370 Wrocław, Poland
[2]Department of Physics, Beijing Normal University,
Beijing 100875, China

Abstract

Superconductivity in a system of strongly correlated electrons treated as a singular Fermi liquid (SFL) is studied. In the system the scattering amplitude for particles with the same momenta is divergent. The qualitatively new ground state and other eigenstates for the superconducting phase are found. Thermodynamic functions are derived within the Bogolubov method. Analytical formulae for the free energy as well as the gap equation and the equation for the chemical potential are presented. The gap equation and the equation for the chemical potential are found to be mutually correlated and have to be simultaneously solved in a self-consistent manner. The numerical solution for the case of the s-wave paired SFL is discussed. It is found that the energy gap and the chemical potential are doubly-valued in the subcritical region, however only the upper branches are thermodynamically stable.

According to the common property of temperature-doping phase diagrams for high-T_c superconductors in the underdoped regime, two characteristic temperatures can be identified in the SFL model. The higher one is the temperature T^* at which phase-incoherent Cooper pairs appear and the pseudogap emerges. The other one is T_c, lower than T^*, at which the Cooper pairs become phase-coherent and form a superconducting condensate. The obtained results evidence that the SFL system displays some features of the superconducting state also in the supercritical temperature range, $T_c < T < T^*$: The non-zero value of the energy gap points at a possibility of Cooper pairs formation, although in terms of the free energy the preferred phase is the normal one. Thus, the supercritical region with a non-zero energy gap can be related to the emergence of the pseudogap. The phase transition to the superconducting state is found to be of the first order with a pronounced discontinuity in the entropy. The hysteresis phenomenon in the supercritical temperature range is considered in detail.

*E-mail address: Mateusz.Krzyzosiak@pwr.wroc.pl

In order to study the response of the system to a weak magnetic field acting on both spin and charge degrees of freedom, the SFL model is extended by including the Pauli term and the superflow. In the extended model the eigenstates for the superconducting phase are found. The spin susceptibility and the magnetic field penetration depth are discussed. It is shown that the normal-phase spin susceptibility has a Curie-law character. Critical parameters such as the critical magnetic field and the critical current are studied employing the variational expansion of the free energy.

1. Introduction

In the BCS theory, which achieved a great success in connection and explanation of properties of simple low-T_c superconductors by means of several experimental parameters, the electron system is treated as the Fermi liquid. However, high-T_c materials are more complex, and diverse microscopic correlations, the underlying symmetry and collective effects such as charge density waves and spin density waves compete on their phase diagram. Their normal state near to the optimal doping level shows convincing evidence of being a more composed form of a quantum liquid than the Fermi liquid, and should be considered as a non-Fermi liquid [1–4]. Recently, Varma *et al.* presented a wide class of non-Fermi liquids and showed that some form of a singular Fermi liquid—or marginal Fermi liquid—was observed in copper-oxides in underdoped and overdoped regions near to the optimally doped structure [4].

The Landau theory of the Fermi liquid points to the possibility of its breakdown through instabilities of the collective modes of the Fermi surface, which arise from the solution of the homogeneous part of the Bethe-Salpeter equation. These collective modes can be characterized by the angular momentum l of oscillations of the Fermi surface corresponding to the spin-direct (symmetric in spin) or the spin-exchange (antisymmetric in spin) Fermi-liquid interaction (Landau function). The absence of a divergence in the Landau function is consistent with the Pomeranchuk stability conditions imposed on the Landau parameters. These conditions, which are of the form $G_l^{\mathrm{a,s}} > -1$ and $F_l^{\mathrm{s,a}} > -(2l+1)$ (*cf.* appendix A) are necessary to ensure the stability of the two- and three-dimensional Fermi liquid [4–7]. Violations of whichever one of these conditions are considered as Landau-Pomeranchuk singularities leading to the formation of singular or non-Fermi liquids. The violation of $l = 0$ conditions (*i.e.* $F_0^{\mathrm{s,a}} \leq -1$) refers to the divergence in the compressibility and the uniform spin susceptibility. The former one would in general occur via a first-order transition whereas the latter one describes the ferromagnetic instability. No other Landau-Pomeranchuk instabilities have been experimentally identified. However, such new and exotic possibilities should be kept in mind. Thus, for example, the F_1^{s} instability corresponds to the Fermi velocity tending to 0, the F_2^{s} instability to the "*d*-wave-like" instability of the particle-hole excitations on the Fermi surface *etc.* Presumably, these instabilities are resolved by the reconstruction of the Fermi surface. Microscopic interactions necessary for the Landau-Pomeranchuk instabilities and critical properties near such instabilities have not been well investigated so far, especially for fermions with a lattice potential [4]. Hence, some specific conditions imposed on the Landau parameters can lead to some new forms of quantum liquids.

A common property of all temperature–doping phase diagrams [1,2] in the underdoped region is the presence of two characteristic temperatures: the higher one is the temperature T^* in which phase-incoherent Cooper pairs appear and the pseudogap emerges. The other characteristic temperature is T_c, lower than T^*, at which the Cooper pairs become phase-coherent and form a superconducting condensate. In the pseudogap region $T_c < T < T^*$ the system reveals numerous anomalies observed by a number of spectroscopic techniques such as infrared, Raman, and angle-resolved photoemission spectroscopy (ARPES). All these anomalies evidence that superconductivity does not disappear completely at $T = T_c$, but there exists a temperature range where some of the superconducting properties persist also above T_c [8]. Moreover, the symmetry of the pseudogap and the superconducting gap seems to be the same, and the pseudogap regime can be considered as uniform with the width $(T^* - T_c)$ depending on the doping level. The pseudogap state turns to the superconducting one as soon as the temperature is low enough to make the normal phase energetically unfavorable, *i.e.* at $T = T_c$.

In the present chapter we present a model of a singular Fermi liquid with divergent scattering amplitude for particles (electron or holes) with the same quasi-momenta and opposite spins in a narrow, nearly half-filled conduction band. General considerations are carried out for arbitrary forms of dispersion relation and a separable pairing potential covering completely the conduction band. The resulting modification of the chemical potential is also taken into account [9–13].

In the paper we assume $\hbar = c = k_B = 1$.

2. Hamiltonian

The interaction in the real space in a quasi-two-dimensional model of a quantum liquid where *e.g.* hole-bosons interactions in the charge channel are responsible for pairing of particles, can be taken in the form [14]

$$\mathcal{H}_{\text{int}} = \sum_{\langle ij \rangle\, \sigma} V(\mathbf{R}_i - \mathbf{R}_j)\, \hat{n}_{i,\sigma}\hat{n}_{j,-\sigma}, \tag{1}$$

where $\mathbf{R}_i$ is the coordinate of the i-th site and $\hat{n}_{i,\sigma}$ is the number operator. Assuming unit lattice constants ($a_x = a_y = 1$) the potential $V(\mathbf{r})$ can be written in the form

$$V(\mathbf{r}) = V_L - V_0\, \delta(\mathbf{r}) - V_1\, \delta(|\mathbf{r}| - 1). \tag{2}$$

The term $V_L \geq 0$ is responsible for a repulsive, uniform, long-range interaction between particles with opposite spins σ and $-\sigma$ on the area S, $V_0 \geq 0$ corresponds to the on-site attraction, and V_1 takes into account the nearest neighbor attraction. The first term in Hamiltonian (1) with the potential (2) can be rewritten as

$$\frac{1}{2} S V_L \sum_{i,\sigma} \hat{n}_{i,\sigma} \frac{1}{S} \sum_j \hat{n}_{j,-\sigma},$$

where by virtue of the molecular field approximation $\langle n_{-\sigma} \rangle \equiv \frac{1}{S}\sum_j n_{j,\,-\sigma}$ corresponds to the mean number of particles with the opposite spin $-\sigma$ interacting with another particle with spin σ. Such interaction prefers ferromagnetic alignment of spins for every couple of particles at all sites. Hence—according to the Pauli exclusion principle—each site can be occupied merely by a single particle with spin up or down, what coincides with the so-called *spin-liquid* model [15–27]. The interaction (2) transformed to the reciprocal space takes the form

$$V(\mathbf{k}-\mathbf{k}') = R_{\mathbf{k}}\delta_{\mathbf{k}\,\mathbf{k}'} - V_0 - V_1[\cos(k_1-k_1') + \cos(k_2-k_2')],$$

where $R_{\mathbf{k}} = V_{\mathrm{L}}$. Since the first term energetically discriminates the occupation of states with the same momenta, in the limit $R_{\mathbf{k}} \to \infty$ one obtains a *singular* Fermi liquid. Depending on the relation between V_0 and V_1 the s-wave ($V_0 \gg V_1$) or the d-wave ($V_0 \ll V_1$) symmetry state is realized [28–30].

We should also mention that a similar problem of singular interactions was considered by Anderson [31] who showed that for a two-dimensional system with a Fermi surface, the scattering amplitude can be divergent when $\mathbf{k} = \mathbf{k}'$. However, this result was submitted to a critical examination by Engelbrecht *et al.* [32, 33] who proved that the Landau function for a two-dimensional repulsive Fermi system is free of divergences.

In summary, the singular Fermi liquid is described by the effective Hamiltonian

$$\hat{\mathcal{H}} = \sum_{\mathbf{k}}(\epsilon_{\mathbf{k}}-\mu)(\hat{n}_{\mathbf{k}\uparrow}+\hat{n}_{\mathbf{k}\downarrow}) + \sum_{\mathbf{k}} R_{\mathbf{k}}\hat{n}_{\mathbf{k}\uparrow}\hat{n}_{\mathbf{k}\downarrow} - \eta\sum_{\mathbf{k},\mathbf{k}'}\zeta_{\mathbf{k}}\zeta_{\mathbf{k}'}\hat{a}^{\dagger}_{\mathbf{k}\uparrow}\hat{a}^{\dagger}_{-\mathbf{k}\downarrow}\hat{a}_{-\mathbf{k}'\downarrow}\hat{a}_{\mathbf{k}'\uparrow}, \tag{3}$$

where the operator $\hat{a}^{\dagger}_{\mathbf{k}\sigma}$ creates a particle in the state $|\mathbf{k},\sigma\rangle$, and $\hat{n}_{\mathbf{k}\sigma} = \hat{a}^{\dagger}_{\mathbf{k}\sigma}\hat{a}_{\mathbf{k}\sigma}$ is the number operator. The Hamiltonian (3) for a particular form of the dispersion relation $\epsilon_{\mathbf{k}}$, obtained *e.g.* from the two-dimensional tight-binding band model, and a separable pairing potential $\eta\zeta_{\mathbf{k}}\zeta_{\mathbf{k}'}$ can be referred to both: a system with s- or d-wave pairing in the BCS-type or in the Hubbard model.

3. Bogolubov Method

Properties of the singular Fermi liquid can be thoroughly studied employing the Bogolubov method for the grand-canonical system [34,35]. Note that the effective Hamiltonian (3) can be rewritten as

$$\hat{\mathcal{H}} = \sum_{\mathbf{k}}{}' \hat{\kappa}_{\mathbf{k}} + \eta\langle\hat{B}\rangle^2 - \eta\left(\hat{B}^{\dagger} - \langle\hat{B}\rangle\right)\left(\hat{B} - \langle\hat{B}\rangle\right), \tag{4}$$

where

$$\begin{aligned}
\hat{\kappa}_{\mathbf{k}} &= (\epsilon_{\mathbf{k}}-\mu)(\hat{n}_{\mathbf{k}\uparrow}+\hat{n}_{\mathbf{k}\downarrow}+\hat{n}_{-\mathbf{k}\uparrow}+\hat{n}_{-\mathbf{k}\downarrow}) + R_{\mathbf{k}}(\hat{n}_{\mathbf{k}\uparrow}\hat{n}_{\mathbf{k}\downarrow}+\hat{n}_{-\mathbf{k}\uparrow}\hat{n}_{-\mathbf{k}\downarrow}) \\
&\quad - \eta\langle\hat{B}\rangle\zeta_{\mathbf{k}}(\hat{a}^{\dagger}_{\mathbf{k}\uparrow}\hat{a}^{\dagger}_{-\mathbf{k}\downarrow} + \hat{a}^{\dagger}_{-\mathbf{k}\uparrow}\hat{a}^{\dagger}_{\mathbf{k}\downarrow} + \hat{a}_{-\mathbf{k}\downarrow}\hat{a}_{\mathbf{k}\uparrow} + \hat{a}_{\mathbf{k}\downarrow}\hat{a}_{-\mathbf{k}\uparrow}), \\
\hat{B} &= \sum_{\mathbf{k}}\zeta_{\mathbf{k}}\hat{a}_{-\mathbf{k}\downarrow}\hat{a}_{\mathbf{k}\uparrow}.
\end{aligned} \tag{5}$$

Due to the symmetry $\hat{\kappa}_{\mathbf{k}} = \hat{\kappa}_{-\mathbf{k}}$, the primed sum in (4) runs over a half of the reciprocal space only. Following the Bogolubov method, the energy gap (the order parameter) is defined as

$$\Delta_{\mathbf{k}} = \eta\langle\hat{B}\rangle\zeta_{\mathbf{k}}, \tag{6}$$

where $\langle \ldots \rangle$ denotes the self-consistent thermodynamic averaging procedure. Defining the isotropic, depending only on the temperature, amplitude of the energy gap as $\Delta(T) = \eta\langle \hat{B} \rangle$, the order parameter for the BCS-type and the Hubbard model takes the form $\Delta_{\mathbf{k}} = \Delta(T)$ and $\Delta_{\mathbf{k}} = \Delta(T)\,\zeta_{\mathbf{k}}$, respectively.

The Bogolubov method is based on the construction of the approximating Hamiltonian [34, 35]

$$\hat{\mathcal{H}}^0 = \sum_{\mathbf{k}}{}' \hat{\kappa}_{\mathbf{k}} + \eta \langle \hat{B} \rangle^2, \tag{7}$$

with the corresponding thermodynamic potential

$$\Omega = -\frac{1}{\beta} \ln Z + \eta \langle \hat{B} \rangle^2, \tag{8}$$

where $\beta^{-1} = T$. Noting that $[\hat{\kappa}_{\mathbf{k}}, \hat{\kappa}_{\mathbf{k}'}]_- = 0$, the partition function

$$Z = \mathrm{Tr}\exp\left(-\beta \sum_{\mathbf{k}}{}' \hat{\kappa}_{\mathbf{k}}\right)$$

can be—following the Baker–Hausdorff theorem—rewritten as

$$Z = \prod_{\mathbf{k}}{}' \,\mathrm{Tr}\exp(-\beta \hat{\kappa}_{\mathbf{k}}). \tag{9}$$

The Bogolubov method ensures that the thermodynamic potential (8) corresponding to the approximating Hamiltonian (7) in the thermodynamic limit ($N, V \to \infty$, $N/V =$ const) tends to the thermodynamic potential for the effective Hamiltonian (3), provided the parameter $\langle \hat{B} \rangle$ is chosen in such a way that it minimizes the thermodynamic potential (8). Consequently, $\langle \hat{B} \rangle$ taken into account in a self-consistent manner as

$$\langle \hat{B} \rangle = \frac{\mathrm{Tr}\, \hat{B} e^{-\beta \hat{\mathcal{H}}^0}}{\mathrm{Tr}\, e^{-\beta \hat{\mathcal{H}}^0}},$$

i.e. as the thermodynamic average evaluated with the Hamiltonian (7) satisfies this condition.

4. Partition Function

The partial Hamiltonian $\hat{\kappa}_{\mathbf{k}}$ defined by Eq. (5) includes terms acting on states with the momentum $\mathbf{k}$ as well as $-\mathbf{k}$. Hence it acts in a 16-dimensional space of double states

$$\psi_{\sigma\sigma'}(\mathbf{k}) = \left| \begin{array}{cc} \mathbf{k}, & \sigma \\ -\mathbf{k}, & \sigma' \end{array} \right\rangle, \tag{10}$$

where $\sigma, \sigma' = 0, \uparrow, \downarrow$ or $\uparrow\downarrow$ and its matrix elements evaluated in the basis (10) are presented in Table 1.

Table 1. Matrix elements of the Hamiltonian (5) in the full 16-dimensional space of double states; zero off-diagonal elements are omitted and $\tilde{\epsilon}_{\mathbf{k}} = \epsilon_{\mathbf{k}} - \mu$ [9, 10].

	$0,0$	$\uparrow,0$	$\downarrow,0$	$\uparrow\downarrow,0$	$0,\uparrow$	$\uparrow,\uparrow$	$\downarrow,\uparrow$	$\uparrow\downarrow,\uparrow$	$0,\downarrow$	$\uparrow,\downarrow$	$\downarrow,\downarrow$	$\uparrow\downarrow,\downarrow$	$0,\uparrow\downarrow$	$\uparrow,\uparrow\downarrow$	$\downarrow,\uparrow\downarrow$	$\uparrow\downarrow,\uparrow\downarrow$
$0,0$	0						$-\Delta_{\mathbf{k}}$			$-\Delta_{\mathbf{k}}$						
$\uparrow,0$		$\tilde{\epsilon}_{\mathbf{k}}$						$-\Delta_{\mathbf{k}}$								
$\downarrow,0$			$\tilde{\epsilon}_{\mathbf{k}}$									$-\Delta_{\mathbf{k}}$				
$\uparrow\downarrow,0$				$R_{\mathbf{k}} + 2\tilde{\epsilon}_{\mathbf{k}}$												
$0,\uparrow$					$\tilde{\epsilon}_{\mathbf{k}}$									$-\Delta_{\mathbf{k}}$		
$\uparrow,\uparrow$						$2\tilde{\epsilon}_{\mathbf{k}}$										
$\downarrow,\uparrow$	$-\Delta_{\mathbf{k}}$						$2\tilde{\epsilon}_{\mathbf{k}}$									$-\Delta_{\mathbf{k}}$
$\uparrow\downarrow,\uparrow$		$-\Delta_{\mathbf{k}}$						$R_{\mathbf{k}} + 3\tilde{\epsilon}_{\mathbf{k}}$								
$0,\downarrow$									$\tilde{\epsilon}_{\mathbf{k}}$						$-\Delta_{\mathbf{k}}$	
$\uparrow,\downarrow$	$-\Delta_{\mathbf{k}}$									$2\tilde{\epsilon}_{\mathbf{k}}$						$-\Delta_{\mathbf{k}}$
$\downarrow,\downarrow$											$2\tilde{\epsilon}_{\mathbf{k}}$					
$\uparrow\downarrow,\downarrow$			$-\Delta_{\mathbf{k}}$									$R_{\mathbf{k}} + 3\tilde{\epsilon}_{\mathbf{k}}$				
$0,\uparrow\downarrow$													$R_{\mathbf{k}} + 2\tilde{\epsilon}_{\mathbf{k}}$			
$\uparrow,\uparrow\downarrow$					$-\Delta_{\mathbf{k}}$									$R_{\mathbf{k}} + 3\tilde{\epsilon}_{\mathbf{k}}$		
$\downarrow,\uparrow\downarrow$									$-\Delta_{\mathbf{k}}$						$R_{\mathbf{k}} + 3\tilde{\epsilon}_{\mathbf{k}}$	
$\uparrow\downarrow,\uparrow\downarrow$							$-\Delta_{\mathbf{k}}$			$-\Delta_{\mathbf{k}}$						$2R_{\mathbf{k}} + 4\tilde{\epsilon}_{\mathbf{k}}$

In order to find eigenvalues and eigenvectors of the Hamiltonian (5) and then the partition function (9) and the thermodynamic potential (8), the secular equation for the matrix given in Table 1 has to be solved. In the general case, when $R_{\mathbf{k}} > 0$, the 16-dimensional space splits into four invariant subspaces: for one-dimensional, four two-dimensional and one four-dimensional [36]. However, if $R_{\mathbf{k}} \to \infty$, the diagonal matrix elements dominate for seven states with at least one doubly-occupied singlet configuration

$$\psi_{\uparrow\downarrow\,\sigma'}(\mathbf{k}) = \left| \begin{matrix} \mathbf{k}, & \uparrow\downarrow \\ -\mathbf{k}, & \sigma' \end{matrix} \right\rangle, \qquad \psi_{\sigma\,\uparrow\downarrow}(\mathbf{k}) = \left| \begin{matrix} \mathbf{k}, & \sigma \\ -\mathbf{k}, & \uparrow\downarrow \end{matrix} \right\rangle, \tag{11}$$

where $\sigma, \sigma' = 0, \uparrow, \downarrow$ or $\uparrow\downarrow$. The dominating diagonal matrix elements can be then considered as eigenvalues and the corresponding double states as eigenstates of the Hamiltonian (5). Because these seven eigenvalues tend to infinity as $R_{\mathbf{k}} \to \infty$, they do not contribute to the partition function, hence the corresponding double states can be eliminated $R_{\mathbf{k}} \to \infty$ from further consideration and we can restrict the 16-dimensional Fock space to a 9-dimensional one with the basis (10), where $\sigma, \sigma' = 0, \uparrow$ or $\downarrow$. Consequently, the operator $\hat{\kappa}_{\mathbf{k}}$ acting in this space is of the form

$$\begin{aligned} \hat{\kappa}_{\mathbf{k}} &= (\epsilon_{\mathbf{k}} - \mu)(\hat{n}_{\mathbf{k}\uparrow} + \hat{n}_{\mathbf{k}\downarrow} + \hat{n}_{-\mathbf{k}\uparrow} + \hat{n}_{-\mathbf{k}\downarrow}) \\ &\quad -\Delta_{\mathbf{k}}(\hat{a}^{\dagger}_{\mathbf{k}\uparrow}\hat{a}^{\dagger}_{-\mathbf{k}\downarrow} + \hat{a}^{\dagger}_{-\mathbf{k}\uparrow}\hat{a}^{\dagger}_{\mathbf{k}\downarrow} + \text{H.c.}), \end{aligned} \tag{12}$$

where the definition (6) has been taken into account. Matrix elements of the Hamiltonian (12) are given in Table 2.

From this analysis we can conclude, that in the case of a singular Fermi liquid with a strong interaction discriminating double occupations of Bloch states, the stable state of the system is realized in the reduced (9-dimensional) Fock space.

Table 2. Matrix elements of the Hamiltonian (12) in the reduced 9-dimensional space of double states; zero off-diagonal elements are omitted [9, 10].

	$0,0$	$\uparrow,0$	$\downarrow,0$	$0,\uparrow$	$\uparrow,\uparrow$	$\downarrow,\uparrow$	$0,\downarrow$	$\uparrow,\downarrow$	$\downarrow,\downarrow$
$0,0$	0					$-\Delta_{\mathbf{k}}$		$-\Delta_{\mathbf{k}}$	
$\uparrow,0$		$\tilde{\epsilon}_{\mathbf{k}}$							
$\downarrow,0$			$\tilde{\epsilon}_{\mathbf{k}}$						
$0,\uparrow$				$\tilde{\epsilon}_{\mathbf{k}}$					
$\uparrow,\uparrow$					$2\tilde{\epsilon}_{\mathbf{k}}$				
$\downarrow,\uparrow$	$-\Delta_{\mathbf{k}}$					$2\tilde{\epsilon}_{\mathbf{k}}$			
$0,\downarrow$							$\tilde{\epsilon}_{\mathbf{k}}$		
$\uparrow,\downarrow$	$-\Delta_{\mathbf{k}}$							$2\tilde{\epsilon}_{\mathbf{k}}$	
$\downarrow,\downarrow$									$2\tilde{\epsilon}_{\mathbf{k}}$

4.1. Normal Phase

For the normal phase, *i.e.* when $\eta = 0$ (and consequently $\Delta_{\mathbf{k}} = 0$), the consideration can be reduced to the 3-dimensional space, with the single states $|\mathbf{k}, \sigma\rangle$, where $\sigma = 0, \downarrow$ or $\uparrow$,

Table 3. Eigenvalues and eigenstates of the Hamiltonian (5) in the 9-dimensional space of double states [9,10].

eigenvalue	*eigenstate*
$\tilde{\epsilon}_{\mathbf{k}} - E_{\mathbf{k}}$	$\sqrt{\frac{E_{\mathbf{k}}+\tilde{\epsilon}_{\mathbf{k}}}{2E_{\mathbf{k}}}}\,\psi_{00}(\mathbf{k}) - \frac{\sqrt{2}}{2}\sqrt{\frac{E_{\mathbf{k}}-\tilde{\epsilon}_{\mathbf{k}}}{2E_{\mathbf{k}}}}\,[\psi_{\downarrow\uparrow}(\mathbf{k}) + \psi_{\uparrow\downarrow}(\mathbf{k})]$
$\tilde{\epsilon}_{\mathbf{k}}$	$\psi_{\uparrow 0}(\mathbf{k})$
$\tilde{\epsilon}_{\mathbf{k}}$	$\psi_{\downarrow 0}(\mathbf{k})$
$\tilde{\epsilon}_{\mathbf{k}}$	$\psi_{0\uparrow}(\mathbf{k})$
$2\tilde{\epsilon}_{\mathbf{k}}$	$\psi_{\uparrow\uparrow}(\mathbf{k})$
$\tilde{\epsilon}_{\mathbf{k}} + E_{\mathbf{k}}$	$\sqrt{\frac{E_{\mathbf{k}}-\tilde{\epsilon}_{\mathbf{k}}}{2E_{\mathbf{k}}}}\,\psi_{00}(\mathbf{k}) + \frac{\sqrt{2}}{2}\sqrt{\frac{E_{\mathbf{k}}+\tilde{\epsilon}_{\mathbf{k}}}{2E_{\mathbf{k}}}}\,[\psi_{\downarrow\uparrow}(\mathbf{k}) + \psi_{\uparrow\downarrow}(\mathbf{k})]$
$\tilde{\epsilon}_{\mathbf{k}}$	$\psi_{0\downarrow}(\mathbf{k})$
$2\tilde{\epsilon}_{\mathbf{k}}$	$\frac{\sqrt{2}}{2}\,[\psi_{\downarrow\uparrow}(\mathbf{k}) - \psi_{\uparrow\downarrow}(\mathbf{k})]$
$2\tilde{\epsilon}_{\mathbf{k}}$	$\psi_{\downarrow\downarrow}(\mathbf{k})$

forming the basis. Hence, the partition function for the normal phase of the singular Fermi liquid

$$Z_{\mathrm{N}} = \prod_{\mathbf{k}} \left(1 + 2e^{-\beta\tilde{\epsilon}_{\mathbf{k}}}\right) = \prod_{\mathbf{k}} 2\cosh\left[\frac{1}{2}\beta\left(\tilde{\epsilon}_{\mathbf{k}} - \frac{1}{\beta}\ln 2\right)\right] e^{\frac{1}{2}\beta\left(\tilde{\epsilon}_{\mathbf{k}} - \frac{1}{\beta}\ln 2\right)}. \tag{13}$$

4.2. Superconducting Phase

The diagonalization of the Hamiltonian (12) in the reduced 9-dimensional space with excluded doubly-occupied configurations provides the ground state

$$\sqrt{\frac{E_{\mathbf{k}} + \tilde{\epsilon}_{\mathbf{k}}}{2E^*_{\mathbf{k}}}}\,\psi_{00}(\mathbf{k}) - \frac{\sqrt{2}}{2}\sqrt{\frac{E_{\mathbf{k}} - \tilde{\epsilon}_{\mathbf{k}}}{2E_{\mathbf{k}}}}\,[\psi_{\downarrow\uparrow}(\mathbf{k}) + \psi_{\uparrow\downarrow}(\mathbf{k})]$$

with the energy $\tilde{\epsilon}_{\mathbf{k}} - E_{\mathbf{k}}$, where $E_{\mathbf{k}} = \sqrt{\tilde{\epsilon}^2_{\mathbf{k}} + 2\Delta^2_{\mathbf{k}}}$, and the other eigenvalues and eigenstates (*cf.* Table 3).

Having known the eigenvalues we can find the partition function for the superconducting phase of the singular Fermi liquid

$$Z_{\mathrm{S}} = \prod_{\mathbf{k}} \left[\left(4 + 3e^{-\beta\tilde{\epsilon}_{\mathbf{k}}} + 2\cosh\beta E_{\mathbf{k}}\right) e^{-\beta\tilde{\epsilon}_{\mathbf{k}}}\right]^{\frac{1}{2}}. \tag{14}$$

5. Thermodynamics

Thermodynamic characteristics of the studied model can be derived from the free energy found from the thermodynamic potential (8). A system of N particles at the temperature T is in an equilibrium state if the free energy $F = \Omega + \mu N$ is minimized. In the superconducting phase the free energy depends on the parameter $\langle\hat{B}\rangle$ and the chemical potential μ which in general is temperature-dependent. The equilibrium condition implies

$$\frac{\partial\Omega(\mu, \langle\hat{B}\rangle)}{\partial\mu} + N = 0 \qquad \text{and} \qquad \frac{\partial\Omega(\mu, \langle\hat{B}\rangle)}{\partial\langle\hat{B}\rangle} = 0.$$

From the former equation the mean number (concentration) of particles in the system

$$\langle N\rangle = \frac{1}{\beta}\frac{\partial}{\partial\mu}\ln Z \tag{15}$$

has to be constant and temperature-independent, *i.e.* $\langle N\rangle = N$. Moreover, in further calculations—without loss of generality—we assume the chemical potential in the form $\mu(T) = -\frac{1}{\beta}\ln\Upsilon(T)$. On the other hand, for the superconducting phase, from the latter equation and the definition (6) we obtain the gap equation

$$\Delta_{\mathbf{k}} = \frac{\zeta_{\mathbf{k}}}{2\beta}\frac{\partial}{\partial\langle\hat{B}\rangle}\ln Z. \tag{16}$$

Hence for singular Fermi liquids in the superconducting state, the temperature-dependent chemical potential $\mu_{\mathrm{S}}(T)$ and the energy gap $\Delta_{\mathbf{k}}$ have to be simultaneously derived from Eqs. (15) and (16) in a self-consistent manner.

5.1. Normal Phase

Employing the formulas (13) and (15) as well as the relation between the thermodynamic potential (8) and the free energy we find the concentration of particles

$$N = \frac{1}{2}\sum_{\mathbf{k}}\left[1 - \tanh\left(\frac{1}{2}(\beta\tilde{\epsilon}_{\mathbf{k}} - \ln 2)\right)\right]. \tag{17}$$

and the free energy

$$F_{\mathrm{N}}(T) = -\frac{1}{\beta}\sum_{\mathbf{k}}\ln\left[1 + 2e^{-\beta(\epsilon_{\mathbf{k}}-\mu_{\mathrm{N}}(T))}\right] + \mu_{\mathrm{N}}(T)N. \tag{18}$$

for the normal phase.

5.2. Superconducting Phase

On the other hand Eqs. (14) and (15) allow us to find the concentration of particles in the superconducting phase

$$N = \frac{1}{2}\sum_{\mathbf{k}}\left[1 - \frac{-3e^{-\beta\tilde{\epsilon}_{\mathbf{k}}} + 2\frac{\tilde{\epsilon}_{\mathbf{k}}}{E_{\mathbf{k}}}\sinh\beta E_{\mathbf{k}}}{4 + 3e^{-\beta\tilde{\epsilon}_{\mathbf{k}}} + 2\cosh\beta E_{\mathbf{k}}}\right]. \tag{19}$$

The gap equation for *s*- or *d*-wave pairing derived from Eqs. (14) and (16) is of the form

$$\Delta_{\mathbf{k}} = \eta\zeta_{\mathbf{k}}\sum_{\mathbf{k}'}\frac{\zeta_{\mathbf{k}'}\Delta_{\mathbf{k}'}}{E_{\mathbf{k}'}}\frac{\sinh\beta E_{\mathbf{k}'}}{4 + 3e^{-\beta\tilde{\epsilon}_{\mathbf{k}'}} + 2\cosh\beta E_{\mathbf{k}'}}. \tag{20}$$

In the case of *s*-wave pairing for the BCS-type and the Hubbard model the gap equation reduces to

$$\Delta(T) = \lambda\sum_{\mathbf{k}'}\frac{\Delta(T)}{E}\frac{\sinh\beta E}{4 + 3e^{-\beta(\xi-\mu_{\mathrm{S}}(T))} + 2\cosh\beta E}, \tag{21}$$

$$\Delta_{\mathbf{k}} = -\frac{4}{U}\epsilon_{\mathbf{k}}\sum_{\mathbf{k}'}\frac{\epsilon_{\mathbf{k}'}\Delta_{\mathbf{k}'}}{E_{\mathbf{k}'}}\frac{\sinh\beta E_{\mathbf{k}'}}{4 + 3e^{-\beta\tilde{\epsilon}_{\mathbf{k}'}} + 2\cosh\beta E_{\mathbf{k}'}}, \tag{22}$$

respectively. It is worth to emphasize that the gap equation (21), where $\lambda = \eta\zeta_{\mathbf{k}}\zeta_{\mathbf{k}'}$ is the dimensionless coupling parameter in a factorable BCS-type interaction, $\epsilon_{\mathbf{k}} = v_{\mathrm{F}}(|\mathbf{k}| - k_{\mathrm{F}}) = \xi$, and $E = \sqrt{(\xi - \mu_{\mathrm{S}}(T))^2 + 2\Delta^2(T)}$, includes both the energy gap and the chemical potential. Hence, these two quantities must not be calculated independently, but both equations (19) and (20) have to be solved simultaneously in a self-consistent manner. Due to the temperature dependence of the chemical potential, the gap equation is significantly modified in the full temperature range. Hence, the calculations assuming $\mu_{\mathrm{S}}(T) \equiv 0$ yield correct results for $T = 0$ only [36].

Finally, employing the partition function (14) we find the formula for the free energy of the superconducting phase

$$\begin{aligned} F_{\mathrm{S}}(T) &= -\frac{1}{2\beta}\sum_{\mathbf{k}} \ln\left(4 + 3e^{-\beta(\epsilon_{\mathbf{k}}-\mu_{\mathrm{S}}(T))} + 2\cosh\beta E_{\mathbf{k}}\right) \\ &+\frac{1}{2}\sum_{\mathbf{k}}(\epsilon_{\mathbf{k}} - \mu_{\mathrm{S}}(T)) + \frac{\Delta_{\mathbf{k}}^2}{\eta\zeta_{\mathbf{k}}^2} + \mu_{\mathrm{S}}(T)N. \end{aligned} \tag{23}$$

6. Thermodynamics of *s*-wave Paired Singular Fermi Liquid

In order to reveal fundamental properties of the singular Fermi liquid system and to study the phase transition between the normal and the superconducting phase, in the present section we consider in detail a system with s-wave pairing [9, 10, 12, 13]. We assume that in the system the conduction band is nearly half-filled, and the pairing interaction covers completely the conduction band, so that the cut-off parameters of the pairing interaction are determined by the band boundaries $-\omega$ and ω [37]. The summation over the particle momentum in all formulas is replaced by the integration over its energy by applying the extended Van Hove scenario [38]. Because the considered system is purely s-wave paired, instead of the kernel of the density of states, in all calculations we can employ its angular average, *i.e.* the usual density of states $N(\xi) = \int_0^{2\pi} \frac{\mathrm{d}\varphi}{2\pi}\, \mathcal{K}(\xi, \varphi)$ [37, 38]. Moreover, we assume that the changes of the density of states within the conduction band are negligible, *i.e.* $N(\xi) \approx N(0)$.

6.1. Normal Phase

Analyzing Eq. (17) we find $\Upsilon(T) = 2$, *i.e.* $\mu_{\mathrm{N}}(T) = -\frac{1}{\beta}\ln 2$. Moreover, employing the extended Van Hove Scenario, from Eq. (18) we obtain the formula

$$F_{\mathrm{N}}(T) = -\frac{2}{\beta}N(0)\int_0^{\omega} \ln\left(1 + e^{-\beta\xi}\right)\,\mathrm{d}\xi - \frac{1}{2}N(0)\omega^2 - \frac{1}{\beta}N(0)\omega\ln 2, \tag{24}$$

determining the free energy of the normal phase. Eq. (24) implies, that the residual entropy of the singular Fermi liquid $S_{\mathrm{N}}(0) = N\ln 2$, contrary to the Fermi-liquid value $S_{\mathrm{N}}(0) = 0$. Let us note that in the presence of the residual entropy, the free energy decreases with increasing temperature.

6.2. Superconducting Phase

Employing the constancy of the mean number of particles, Eqs. (14) and (15) as well as the relation $\frac{\partial}{\partial\mu}(\xi-\mu) = -\frac{\partial}{\partial\xi}(\xi-\mu)$ we obtain the equation

$$0 = -\frac{1}{2\beta}\, N(0) \int_{-\omega}^{\omega} \frac{\partial}{\partial\xi} \ln\left[4 + \frac{3}{\Upsilon(T)}\, e^{-\beta\xi} + 2\cosh\beta E\right] \mathrm{d}\xi.$$

This formula can be rewritten in a compact form

$$\Upsilon(T)^{1+\frac{\omega}{\Sigma}} - \Upsilon(T)^{1-\frac{\omega}{\Sigma}} = 3\,\frac{\sinh\beta\omega}{\sinh\beta\Sigma}, \tag{25}$$

where $\Sigma = \frac{1}{2}\left[\sqrt{(\omega + \frac{1}{\beta}\ln\Upsilon(T))^2 + 2\Delta^2(T)} + \sqrt{(\omega - \frac{1}{\beta}\ln\Upsilon(T))^2 + 2\Delta^2(T)}\right]$. Note that Eq. (25) yields $\Upsilon(T) = 2$ for $\Delta(T) = 0$ and $\Upsilon(0) = 1$ for $\Delta(0) > 0$.

The gap equation (21), where the summation over the momentum has been replaced by the integration over the particle energy according to the extended Van Hove scenario and taking into account that $N(\xi) \approx N(0)$, transforms into

$$1 = \lambda N(0) \int_{-\omega}^{\omega} \frac{\sinh\beta E}{4 + \frac{3}{\Upsilon(T)}\, e^{-\beta\xi} + 2\cosh\beta E}\, \frac{\mathrm{d}\xi}{E}, \tag{26}$$

where $E = \sqrt{\left(\xi + \frac{1}{\beta}\ln\Upsilon(T)\right)^2 + 2\Delta^2(T)}$. The coupling parameter λ is being eliminated in a standard way by employing the gap equation at $T = 0$

$$\frac{1}{\lambda} = \frac{N(0)}{2\Delta^2(0)} \int_{-\omega}^{\omega} \frac{\mathrm{d}\xi}{\sqrt{\xi^2 + 2\Delta^2(0)}}.$$

The superconducting phase of the singular Fermi liquid is completely characterized by two quantities: the chemical potential and the energy gap calculated simultaneously in a self-consistent manner. In order to state whether the superconducting phase is the stable phase realized by the system at a given temperature, we have to compare the corresponding free energy

$$\begin{aligned} F_{\mathrm{S}}(T) \;=\; & -\frac{1}{2\beta}\, N(0) \int_{-\omega}^{\omega} \ln\left(1 + 4e^{-\beta E} + \frac{3}{\Upsilon(T)}\, e^{-\beta(E+\xi)} + e^{-2\beta E}\right) \mathrm{d}\xi \\ & -\frac{1}{4}\, N(0) \left(\omega + \frac{1}{\beta}\ln\Upsilon(T)\right) \sqrt{\left(\omega + \frac{1}{\beta}\ln\Upsilon(T)\right)^2 + 2\Delta^2(T)} \\ & -\frac{1}{4}\, N(0) \left(\omega - \frac{1}{\beta}\ln\Upsilon(T)\right) \sqrt{\left(\omega - \frac{1}{\beta}\ln\Upsilon(T)\right)^2 + 2\Delta^2(T)} \\ & -N(0)\Delta^2(T) \ln\left\{\frac{\Delta(0)}{\Delta(T)\left(\omega + \sqrt{\omega^2 + 2\Delta^2(0)}\right)}\right. \end{aligned}$$

$$\times \left[\omega + \frac{1}{\beta}\ln\Upsilon(T) + \sqrt{\left(\omega + \frac{1}{\beta}\ln\Upsilon(T)\right)^2 + 2\Delta^2(T)}\right]^{\frac{1}{2}}$$

$$\times \left[\omega - \frac{1}{\beta}\ln\Upsilon(T) + \sqrt{\left(\omega - \frac{1}{\beta}\ln\Upsilon(T)\right)^2 + 2\Delta^2(T)}\right]^{\frac{1}{2}}\Bigg\}, \quad (27)$$

found from the formula (23), with the free energy of the normal phase (24). At zero temperature Eq. (27) reduces to

$$F_{\mathrm{S}}(0) = -\frac{1}{2}N(0)\omega\sqrt{\omega^2 + 2\Delta^2(0)}.$$

Taking into account the form of the free energy for the normal phase (24), we find the free energy difference at $T = 0$

$$F_{\mathrm{S}}(0) - F_{\mathrm{N}}(0) = -\frac{1}{2}N(0)\left(\omega\sqrt{\omega^2 + 2\Delta^2(0)} - \omega^2\right), \quad (28)$$

which is presented in Figure 1 for various values of the conduction band width. In the case

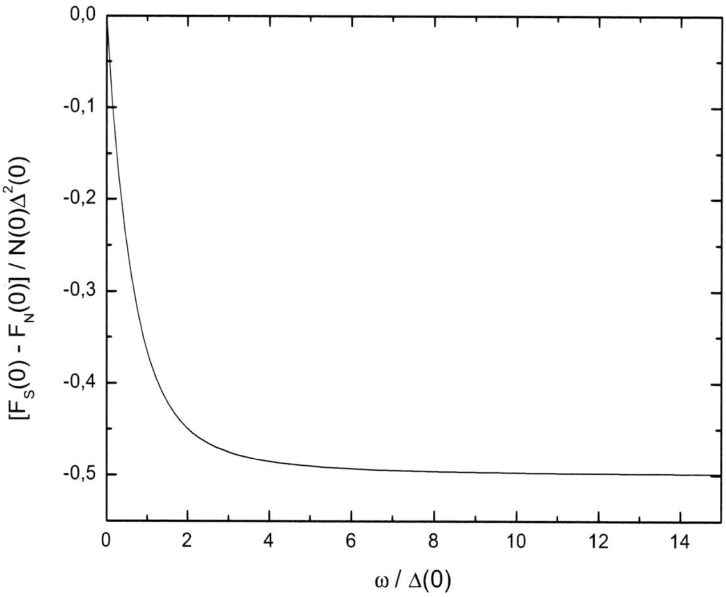

Figure 1. The free energy difference between the superconducting and the normal phase (28) for the singular Fermi liquid at $T = 0$ for various values of the conduction band width.

of a wide conduction band, *i.e.* $\omega \gg \Delta(0)$, we have

$$F_{\mathrm{S}}(0) - F_{\mathrm{N}}(0) = -\frac{1}{2}N(0)\Delta^2(0),$$

which is the standard BCS model result.

Although the presented Bogolubov method allows us to derive the complete form of the free energy for the superconducting and the normal phase, in many other approaches the free energy difference between the superconducting and the normal phase $\Delta F(T) = F_{\mathrm{S}}(T) - F_{\mathrm{N}}(T)$ is reproduced from the energy gap [37]. If the chemical potentials for both phases are identical, then $\Delta F(T) = \Delta\Omega(T)$, and the differences of the entropy and specific heat can be found as

$$\Delta S(T) = -\frac{\mathrm{d}\Delta\Omega(T)}{\mathrm{d}T}, \qquad \Delta C(T) = -T\,\frac{\mathrm{d}^2\Delta\Omega(T)}{\mathrm{d}T^2}. \tag{29}$$

However, when the chemical potentials for both phases do differ, then $\Delta\Omega(T) \neq \Delta F(T)$ and $\Delta S(T)$ as well as $\Delta C(T)$ must be derived from the free energy difference. In the large ω limit [37]

$$\Delta F(T) = \int_0^{\lambda} \Delta^2(T)\,\frac{\mathrm{d}}{\mathrm{d}\lambda'}\,\frac{1}{\lambda'}\,\mathrm{d}\lambda' - \frac{1}{2}\,N(0)\left[T\ln\frac{1}{2}\Upsilon(T)\right]^2. \tag{30}$$

Hence the formulas for the entropy and the specific heat of the superconducting phase have to contain extra terms due to the chemical potential contribution. If for any other models (*e.g.* when the conduction band is not half-filled) the function Υ is almost temperature-independent, then the contribution to the superconducting specific heat corresponding to the latter term in Eq. (30) is equal to $N(0)T\left(\ln\frac{1}{2}\Upsilon\right)^2$. Since the specific heat of the normal phase $C_{\mathrm{N}}(T) = \frac{\pi^2}{3}\,N(0)T$, we can assume that the change in the chemical potential of the superconducting phase corresponds to the change of the effective mass in the normal phase, which is modified according to the relation $\frac{m^*}{m} = 1 + \frac{3}{\pi^2}\left(\ln\frac{1}{2}\Upsilon\right)^2$ [39]. Then the specific heat of the superconducting phase can be found by calculating only the integral in the formula (30). On the other hand, in the case of the discussed singular Fermi liquid, the latter term in Eq. (30) can be approximated by $-\frac{1}{2}\,N(0)(T\ln 2)^2$. Hence, the corresponding contribution to the specific heat is $N(0)T(\ln 2)^2$, which is $0.146\,C_{\mathrm{N}}(T)$. The above discussion proves that the temperature dependence of the energy gap does not carry sufficiently enough information in order to allow us to reconstruct the thermodynamic properties of the system. Therefore—in general—it has to be supplemented by the chemical potential. Thus, when considering superconductivity in strongly correlated electron systems with a narrow band and relatively strong coupling, the applied approaches should be based on the many-body renormalization procedure. Moreover, the chemical potential as a function of both the temperature and the particle concentration has to be independently calculated for the normal and for the superconducting phase.

More detailed investigations of thermodynamic functions can be performed employing some universal relations between the free energy, the entropy, the specific heat difference and the energy gap amplitude, which have been established independently [37]. These relations can be applied to a wide class of models of superconductivity. Such results as well as some previous ones [40], obtained under more general conditions to be satisfied by a superconducting system, should turn out to be useful in explanation of common elements of the mechanism of low- and high-T_{c} superconductivity.

7. Numerical Results and Discussion

In order to illustrate the fundamental properties of the studied model in more detail, the analytical results obtained in the previous section for the s-wave paired singular Fermi liquid have been supplemented by numerical calculations. The comparison of various quantities measured in the energy units will be facilitated if they are normalized by $\Delta(0)$, *i.e.* the zero-temperature value of the energy gap. Hence, in further calculations we consequently employ their dimensionless counterparts. The numerical results presented in this section have been obtained for three systems with the conduction band of a small, medium and large width ($\omega/\Delta(0) = 2, 2\sqrt{5}$ and 10, respectively), allowing us to identify the universal properties of the model.

In Figures 2a and 3a the self-consistent solutions of the gap equation (26) and the equation for the chemical potential (25), *i.e.* $\Delta(T)/\Delta(0)$ and $\mu(T)/\Delta(0)$, are presented. Additionally in Figure 4 the temperature dependence of the parameter Υ is plotted.

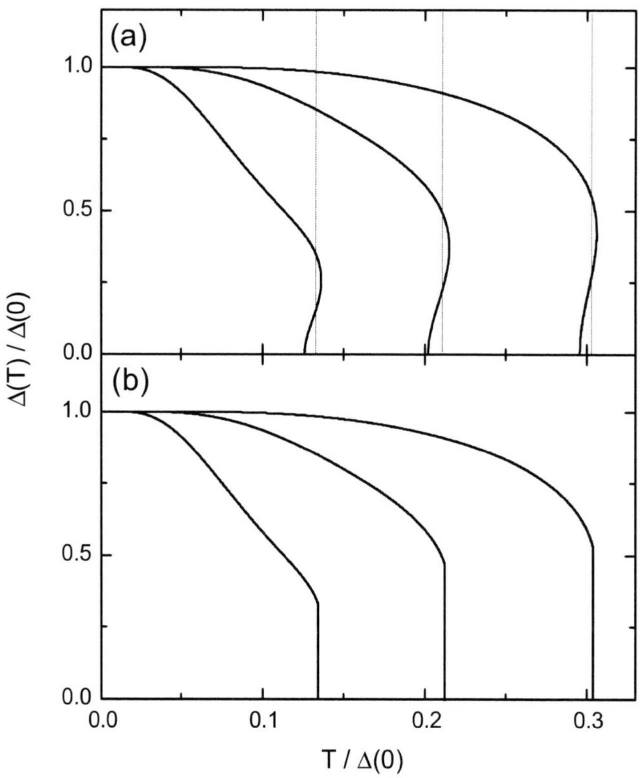

Figure 2. The energy gap in the singular Fermi liquid model for $\omega/\Delta(0) = 2$, $2\sqrt{5}$ and 10 (curves from the right to the left): (a) full solutions including both branches, (b) only the stable branch is plotted: the phase transition to the normal phase at T_c was taken into account. The dotted vertical lines mark the phase transition temperature T_c [9, 10, 12].

Analyzing the results presented in Figures 2a and 3a we can conclude that in the temperature range $0 < T < T_0$ the gap equation (and the equation for the chemical potential,

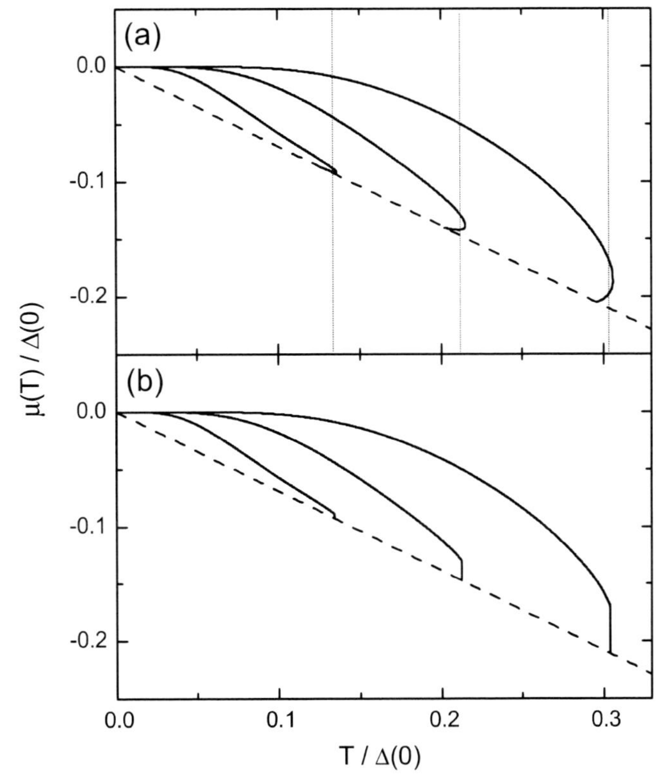

Figure 3. The chemical potential in the singular Fermi liquid model for $\omega/\Delta(0) = 2, 2\sqrt{5}$ and 10 (curves from the right to the left): (a) full solutions including both branches, (b) only the stable branch is plotted: the phase transition to the normal phase at T_c was taken into account. The dotted vertical lines mark the phase transition temperature T_c [9, 10, 12].

consequently) has one solution. At the temperature $T = T_0$ another solution emerges, opening the lower branch of the energy gap (chemical potential). Denoting by T^* the highest temperature at which a non-zero solution of the gap (the chemical potential) equation exist we state that in the temperature range $T_0 < T < T^*$ the energy gap equation (and the equation for the chemical potential) has two solutions.

In order to resolve for which of these solutions the corresponding free energy is lower than the free energy of the normal phase (*i.e.* which solution and in what temperature range is realized in the system being in the superconducting state), the values of the free energy corresponding to both solutions have been calculated (*cf.* Eq. (27)) and compared with the free energy of the normal phase (24).

The results presented in Figure 5 allow us to conclude that the free energy corresponding to the lower branch of the energy gap (the chemical potential) is—in the full temperature range $T_0 < T < T^*$—greater than the free energy of the normal phase. On the other hand, the free energy of the superconducting phase corresponding to the higher branch is less than the free energy of the normal phase for $0 \leq T < T_c$. For $T_c < T \leq T^*$ the free energy of the normal phase is the lowest one and the system stops to be superconducting.

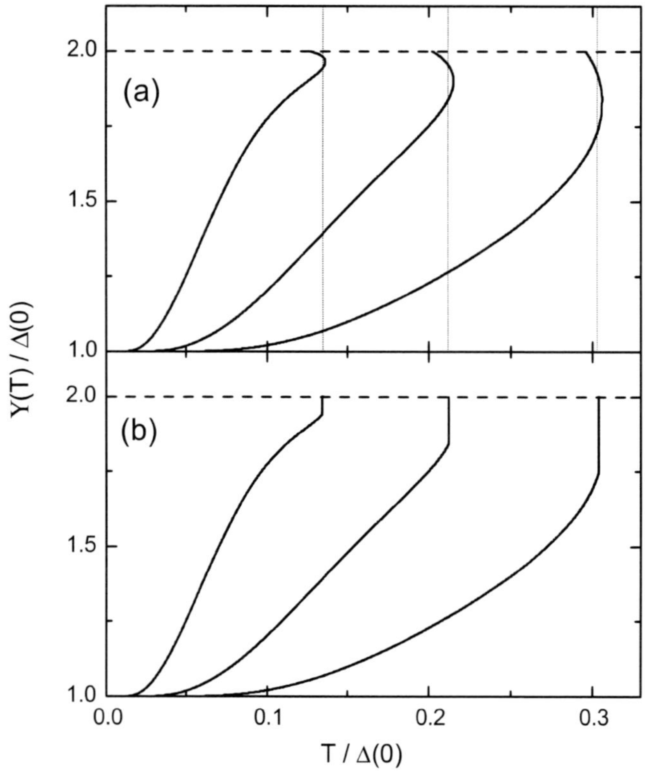

Figure 4. The parameter Υ in the singular Fermi liquid model for $\omega/\Delta(0) = 2$, $2\sqrt{5}$ and 10 (curves from the right to the left): (a) full solutions including both branches, (b) only the stable branch is plotted: the phase transition to the normal phase at T_c was taken into account. The dotted vertical lines mark the phase transition temperature T_c [9, 10, 12].

The temperature T_c, where the values of the free energy for both phases are equal (*i.e.* $\Delta F(T_c) = 0$), is the temperature of the phase transition from the superconducting to the normal phase. Taking into account the phase transition at $T = T_c$, the forms of the energy gap, the chemical potential and the parameter Υ have been accordingly modified: in Figures 2b, 3b and 4b we present only the solutions realized in the system, *i.e.* the solutions chosen on the free energy criterion. These solutions correspond to the upper branch of the energy gap (the chemical potential) for $0 \leq T \leq T_c$.

The presented results evidence that the system displays some features of the superconducting state also in the supercritical temperature range, $T_c < T \leq T^*$. The non-zero value of the energy gap points at a possibility of Cooper pairs formation, although in terms of the free energy the preferred phase is the normal one. The supercritical region with a non-zero energy gap (*cf.* Figures 2a and 5c) can be related to the emergence of the pseudogap. In the superconducting Fermi liquid system the following scenario can be realized: At the temperature T^* phase-incoherent Cooper pairs can be formed and the pseudogap opens. As a result of lowering the temperature further down to T_c—when the normal phase becomes (in terms of the free energy) disadvantageous for the system—the preformed Cooper pairs gain

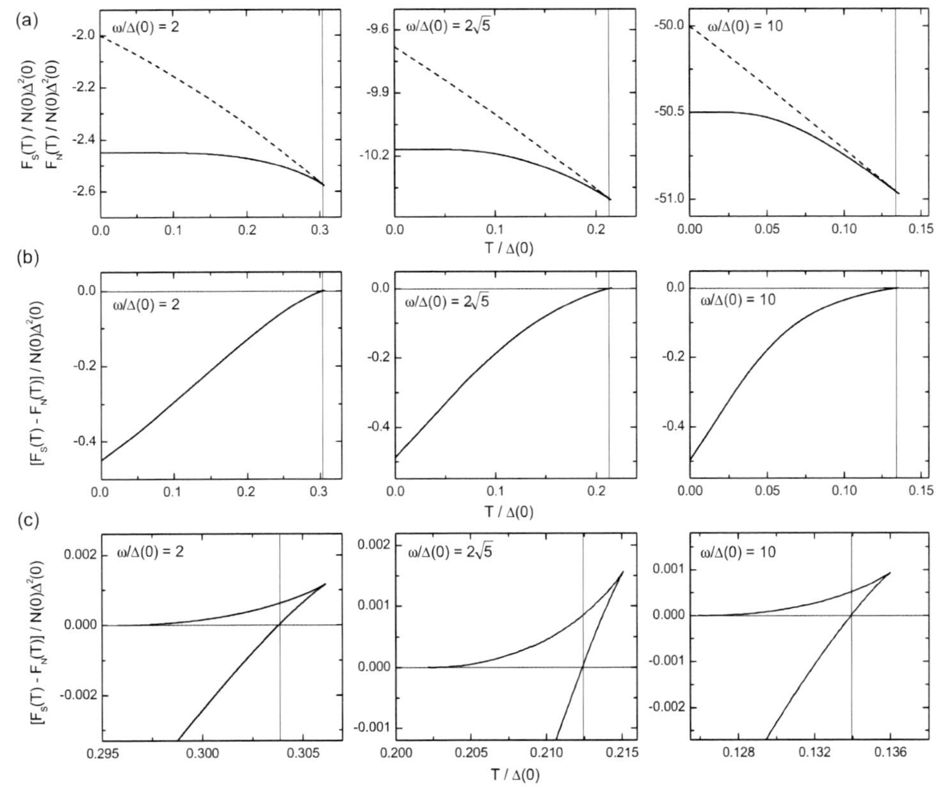

Figure 5. The temperature dependence of: (a) the free energy of the superconducting phase (solid) and the normal phase (dashed), (b) the difference of the free energy values, (c) the difference of the free energy values in the critical region, for the singular Fermi liquid. The dotted vertical lines mark the phase transition temperatures T_c [9, 10, 12, 13].

the phase coherence and the systems undergoes a phase transition to the superconducting phase. Because for $T_c < T \leq T^*$ the free energy corresponding to the higher branch of the energy gap (and the chemical potential) is higher than the free energy of the normal phase, in this temperature range the superconducting state can be only metastable (*cf.* Figure 6). It implies that a *gap-like* behavior is observed in ARPES and many other experiments done on high-T_c materials. Moreover, estimations of the order parameter obtained by means of experimental observations reveal significant differences in its value when approaching the superconducting transition form above (decreasing temperature) and below (increasing temperature). This testifies to the hysteretic behavior of high-T_c superconductors for $T_c < T < T^*$ [8]. So, the crossover scheme with the related pseudogap state displayed below T^* in the present formalism clearly corresponds to the metastable superconducting state region with the boundary delineated by the spinodal.

The values of the three characteristic temperatures, fulfilling the relation $T_0 < T_c < T^*$, are given in Table 4. Moreover, in Tables 5 and 6 values of some selected thermodynamic characteristics at the pseudogap-opening temperature T^* and the phase transition temperature T_c are quoted.

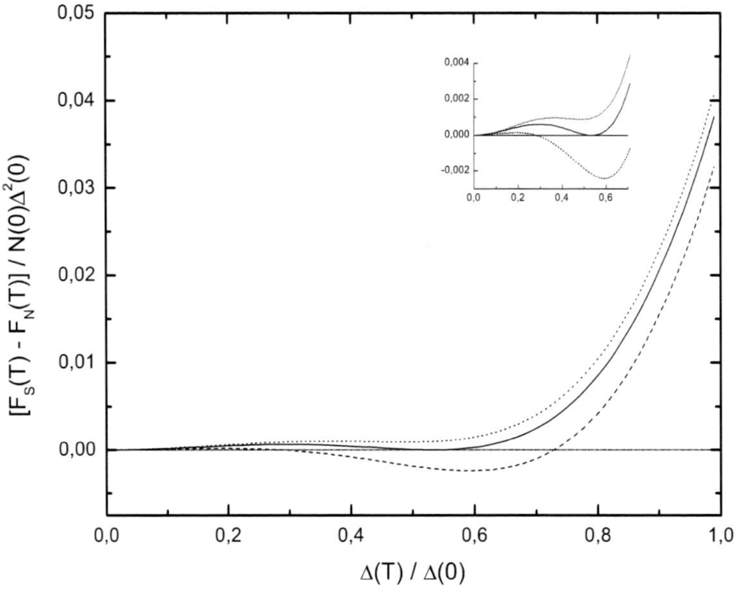

Figure 6. The free energy difference as a function of the energy gap for $\omega/\Delta(0) = 2$. For each given value of the energy gap the chemical potential has been calculated. The calculations have been performed for three temperature values: $T_c/\Delta(0) = 0.3038$ (solid), $T_1/\Delta(0) = 0.3$ (dashed) oraz $T_2/\Delta(0) = 0.3055$ (dotted) fulfilling the relation $T_0 < T_1 < T_c < T_2 < T^*$ For the temperature T_2 a metastable state in the pseudogap region is possible. The horizontal line indicates the zero level.

Table 4. The characteristic temperatures in the superconducting Fermi liquid model.

$\frac{\omega}{\Delta(0)}$	$\frac{T_0}{\Delta(0)}$	$\frac{T_c}{\Delta(0)}$	$\frac{T^*}{\Delta(0)}$	$\frac{\Delta(0)}{T_0}$	$\frac{\Delta(0)}{T_c}$	$\frac{\Delta(0)}{T^*}$
2	0.2958	0.3038	0.3061	3.381	3.292	3.267
$2\sqrt{5}$	0.2022	0.2124	0.2151	4.946	4.708	4.649
10	0.1259	0.1339	0.1359	7.943	7.469	7.358

Table 5. Thermodynamic characteristics of the model at the pseudogap-opening temperature T^*.

$\frac{\omega}{\Delta(0)}$	$\frac{T^*}{\Delta(0)}$	$\frac{\Delta(T^*)}{\Delta(0)}$	$\frac{\mu(T^*)}{\Delta(0)}$	$\Upsilon(T^*)$	$\frac{\Delta F(T^*)}{N(0)\Delta^2(0)}$
2	0.3061	0.4403	-0.1844	1.8266	$1.1653 \cdot 10^{-3}$
$2\sqrt{5}$	0.2151	0.3706	-0.1383	1.9017	$1.5672 \cdot 10^{-3}$
10	0.1359	0.2585	-0.0917	1.9639	$0.9377 \cdot 10^{-3}$

In order to investigate the nature of the phase transition—which already from Figure

5c can be identified as a first-order one—in more detail, the calculated values of the free energy have been employed to find the entropy $S(T) = -\mathrm{d}F(T)/\mathrm{d}T$ and the specific heat $C(T) = -T\mathrm{d}^2F(T)/\mathrm{d}T^2$ for both phases in the temperature range $0 < T < T_\mathrm{c}$. The temperature dependence of the entropy (*cf.* Figure 7) reveals a pronounced discontinuity confirming that the phase transition is of the first order with a non-zero heat of the transition (*cf.* Table 6). ¿From Figures 8 and 9 we can note that the presented temperature dependence

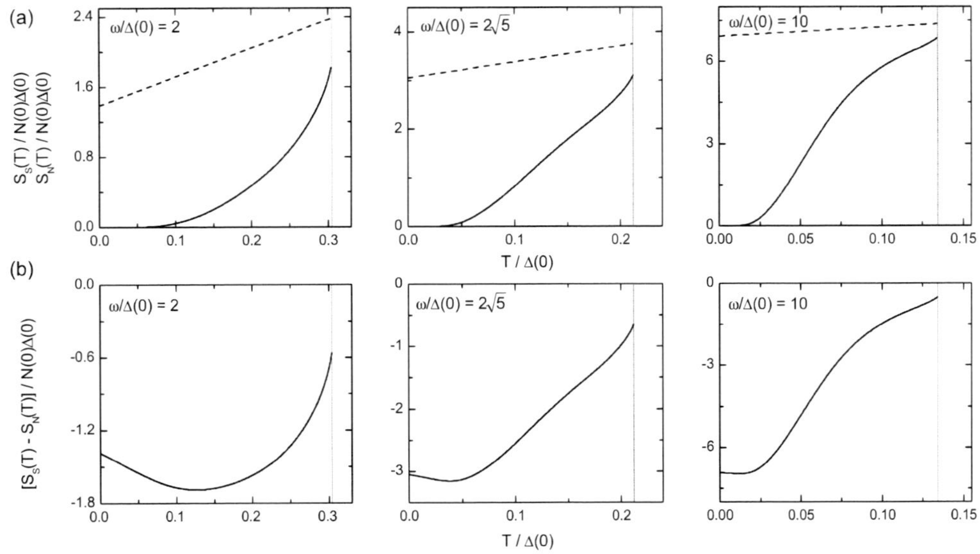

Figure 7. The temperature dependence of (a) the superconducting-phase entropy (solid) and the normal-phase entropy (dashed), (b) the entropy difference. The dotted vertical lines mark the phase-transition temperature T_c [9, 10].

Table 6. Thermodynamic characteristics of the model at the phase transition temperature T_c.

$\frac{\omega}{\Delta(0)}$	$\frac{T_\mathrm{c}}{\Delta(0)}$	$\frac{\Delta(T_\mathrm{c})}{\Delta(0)}$	$\frac{\mu(T_\mathrm{c})}{\Delta(0)}$	$\Upsilon(T_\mathrm{c})$	$\frac{\Delta S(T_\mathrm{c})}{N(0)\Delta(0)}$	$\frac{\Delta C(T_\mathrm{c})}{N(0)\Delta(0)}$	$\frac{T_\mathrm{c}\Delta S(T_\mathrm{c})}{N(0)\Delta^2(0)}$
2	0.3038	0.5320	−0.1691	1.7450	0.5659	13.25	0.1719
$2\sqrt{5}$	0.2124	0.4704	−0.1301	1.8450	0.6498	9.27	0.1380
10	0.1339	0.3324	−0.0888	1.9411	0.5150	6.41	0.0689

of the specific heat is strongly modified with as the conduction band width changes. For a system with a wide band we observe an extremum (anomaly) of the specific heat, whereas for systems with a narrow band these anomalies are not observed (*cf.* Figure 8).

The dependence of the temperature T_0 at which the lower branch of the energy gap opens and the phase transition temperature T_c on the band-width up to around $\omega/\Delta(0) = 20$ is presented in Figure 10.

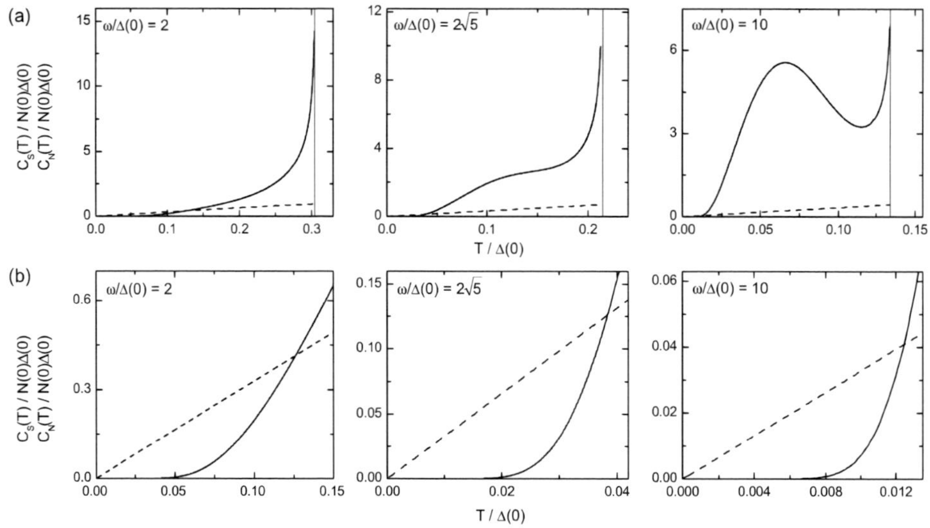

Figure 8. The specific heat of the superconducting (solid) and the normal (dashed) phase: (a) in the full temperature range, (b) in the critical temperature region. The dotted vertical lines mark the phase transition temperature T_c [9, 10].

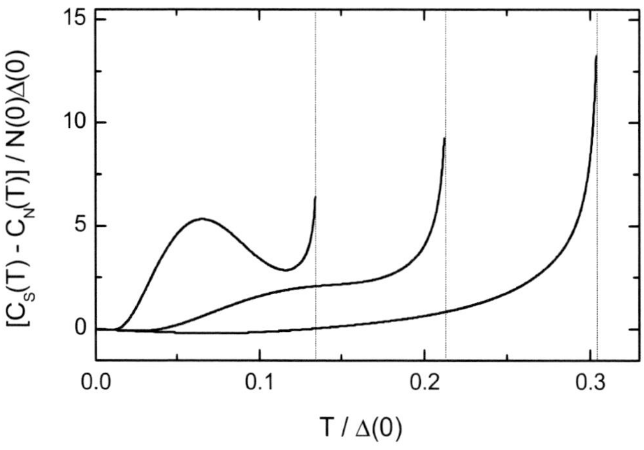

Figure 9. The specific heat difference for $\omega/\Delta(0) = 2, 2\sqrt{5}$ and 10 (curves from the right to the left). The dotted vertical lines mark the phase transition temperature T_c [9, 10].

8. Response to External Magnetic Field

The discussion presented in previous sections referred to the singular Fermi liquid model in the absence of any external fields or currents. In this section we present studies of the system's response to a weak magnetic field interacting with the spin and the charge degrees of freedom [11, 12], as well as a discussion of critical parameters related to the magnetic field [11, 13]. The numerical calculations have been performed for the singular Fermi liquid

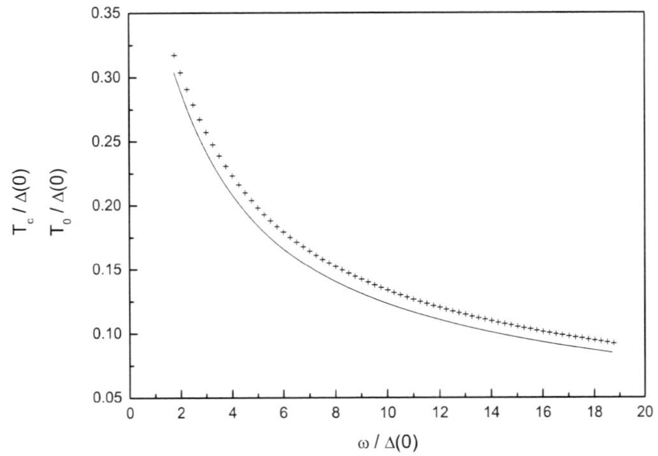

Figure 10. The phase transition temperature T_c (crosses) and the temperature T_0 (solid) vs. band-width.

model with purely s-wave pairing defined in Section 7..

8.1. Spin Susceptibility

The interaction of spin degrees of freedom with an external magnetic field $\mathbf{H}$ enters the Hamiltonian by means of the Pauli term $-\mu_B Hs$, where s is the projection of the spin on the field direction and μ_B is the Bohr magneton. In order to study the effects implied by this interaction, we therefore have to consider the following Hamiltonian [11]

$$\begin{aligned} \hat{\kappa}_{\mathbf{k}} &= (\epsilon_{\mathbf{k}} - \mu)(\hat{n}_{\mathbf{k}\uparrow} + \hat{n}_{\mathbf{k}\downarrow} + \hat{n}_{-\mathbf{k}\uparrow} + \hat{n}_{-\mathbf{k}\downarrow}) - h(\hat{n}_{\mathbf{k}\uparrow} - \hat{n}_{\mathbf{k}\downarrow} + \hat{n}_{-\mathbf{k}\uparrow} - \hat{n}_{-\mathbf{k}\downarrow}) \\ &\quad -\Delta_{\mathbf{k}}(\hat{a}^\dagger_{\mathbf{k}\uparrow}\hat{a}^\dagger_{-\mathbf{k}\downarrow} + \hat{a}^\dagger_{-\mathbf{k}\uparrow}\hat{a}^\dagger_{\mathbf{k}\downarrow} + \text{H.c.}), \end{aligned} \tag{31}$$

where $h = \mu_B H$ jest the magnitude of the magnetic field expressed in the energy units. Note that taking the Pauli interaction into account does not require to change the basis (10) with $\sigma, \sigma' = 0, \uparrow$ or $\downarrow$. Therefore we can easily find the matrix elements of the modified Hamiltonian (*cf.* Table 7).

The diagonalization of the Hamiltonian (31) provides eigenvalues and eigenvectors (*cf.* Table 8) and allows us to calculate the partition function for the superconducting phase

$$Z_S = \prod_{\mathbf{k}} \left\{ \left[4\cosh\beta h + (1 + 2\cosh 2\beta h)\, e^{-\beta\tilde{\epsilon}_{\mathbf{k}}} + 2\cosh\beta E_{\mathbf{k}} \right] e^{-\beta\tilde{\epsilon}_{\mathbf{k}}} \right\}^{\frac{1}{2}}. \tag{32}$$

On the other hand, the partition function for the normal phase, *i.e.* when $\Delta_{\mathbf{k}} = 0$, reads

$$Z_N = \prod_{\mathbf{k}} \left[1 + 2e^{-\beta\tilde{\epsilon}_{\mathbf{k}}} \cosh\beta h \right]. \tag{33}$$

Table 7. Matrix elements of the Hamiltonian (31); off-diagonal zeros omitted [11].

	$0,0$	$\uparrow,0$	$\downarrow,0$	$0,\uparrow$	$\uparrow,\uparrow$	$\downarrow,\uparrow$	$0,\downarrow$	$\uparrow,\downarrow$	$\downarrow,\downarrow$
$0,0$	0					$-\Delta_{\mathbf{k}}$		$-\Delta_{\mathbf{k}}$	
$\uparrow,0$		$\tilde{\epsilon}_{\mathbf{k}}-h$							
$\downarrow,0$			$\tilde{\epsilon}_{\mathbf{k}}+h$						
$0,\uparrow$				$\tilde{\epsilon}_{\mathbf{k}}-h$					
$\uparrow,\uparrow$					$2\tilde{\epsilon}_{\mathbf{k}}-2h$				
$\downarrow,\uparrow$	$-\Delta_{\mathbf{k}}$					$2\tilde{\epsilon}_{\mathbf{k}}$			
$0,\downarrow$							$\tilde{\epsilon}_{\mathbf{k}}+h$		
$\uparrow,\downarrow$	$-\Delta_{\mathbf{k}}$							$2\tilde{\epsilon}_{\mathbf{k}}$	
$\downarrow,\downarrow$									$2\tilde{\epsilon}_{\mathbf{k}}+2h$

Table 8. Eigenvalues and eigenstates of the Hamiltonian (31) [11].

eigenvalue	*eigenvector*
$\tilde{\epsilon}_{\mathbf{k}}-E_{\mathbf{k}}$	$\sqrt{\frac{E_{\mathbf{k}}+\tilde{\epsilon}_{\mathbf{k}}}{2E_{\mathbf{k}}}}\,\psi_{00}(\mathbf{k})-\frac{\sqrt{2}}{2}\sqrt{\frac{E_{\mathbf{k}}-\tilde{\epsilon}_{\mathbf{k}}}{2E_{\mathbf{k}}}}\,[\psi_{\downarrow\uparrow}(\mathbf{k})+\psi_{\uparrow\downarrow}(\mathbf{k})]$
$\tilde{\epsilon}_{\mathbf{k}}-h$	$\psi_{\uparrow 0}(\mathbf{k})$
$\tilde{\epsilon}_{\mathbf{k}}+h$	$\psi_{\downarrow 0}(\mathbf{k})$
$\tilde{\epsilon}_{\mathbf{k}}-h$	$\psi_{0\uparrow}(\mathbf{k})$
$2\tilde{\epsilon}_{\mathbf{k}}-2h$	$\psi_{\uparrow\uparrow}(\mathbf{k})$
$\tilde{\epsilon}_{\mathbf{k}}+E_{\mathbf{k}}$	$\sqrt{\frac{E_{\mathbf{k}}-\tilde{\epsilon}_{\mathbf{k}}}{2E_{\mathbf{k}}}}\,\psi_{00}(\mathbf{k})+\frac{\sqrt{2}}{2}\sqrt{\frac{E_{\mathbf{k}}+\tilde{\epsilon}_{\mathbf{k}}}{2E_{\mathbf{k}}}}\,[\psi_{\downarrow\uparrow}(\mathbf{k})+\psi_{\uparrow\downarrow}(\mathbf{k})]$
$\tilde{\epsilon}_{\mathbf{k}}+h$	$\psi_{0\downarrow}(\mathbf{k})$
$2\tilde{\epsilon}_{\mathbf{k}}$	$\frac{\sqrt{2}}{2}\,[\psi_{\downarrow\uparrow}(\mathbf{k})-\psi_{\uparrow\downarrow}(\mathbf{k})]$
$2\tilde{\epsilon}_{\mathbf{k}}+2h$	$\psi_{\downarrow\downarrow}(\mathbf{k})$

The magnetization $M(T,h)$ and the spin susceptibility $\chi(T)$ are related to the partition function by the standard formulas

$$M(T,h)=\frac{\mu_{\mathrm{B}}}{\beta}\frac{\partial\ln Z}{\partial h},\qquad \chi(T)=\mu_{\mathrm{B}}\left.\frac{\partial M(T,h)}{\partial h}\right|_{h=0}.$$

Hence, employing the calculated partition function (32) one obtains the magnetization and the spin susceptibility of the superconducting phase

$$\begin{aligned} M_{\mathrm{S}}(T,h) &= 2\mu_{\mathrm{B}}\sum_{\mathbf{k}}\frac{\sinh\beta h+e^{-\beta\tilde{\epsilon}_{\mathbf{k}}}\sinh 2\beta h}{4\cosh\beta h+(1+2\cosh 2\beta h)\,e^{-\beta\tilde{\epsilon}_{\mathbf{k}}}+2\cosh\beta E_{\mathbf{k}}}, \\ \chi_{\mathrm{S}}(T) &= 2\mu_{\mathrm{B}}^{2}\beta\sum_{\mathbf{k}}\frac{1+2e^{-\beta\tilde{\epsilon}_{\mathbf{k}}}}{4+3e^{-\beta\tilde{\epsilon}_{\mathbf{k}}}+2\cosh\beta E_{\mathbf{k}}}. \end{aligned}$$

Since $\cosh\beta E_{\mathbf{k}}\to\infty$ as $T\to 0$, the spin susceptibility of the superconducting phase vanishes at zero temperature, *i.e.* $\chi_{\mathrm{S}}(0)=0$.

Further, we replace the summation over the particle momentum by the integration over the particle energy according to the extended Van Hove Scenario, obtaining eventually in the case of the singular Fermi liquid with purely s-wave pairing and $N(\xi) \approx N(0)$

$$\chi_{\rm S}(T) = 2\mu_{\rm B}^2\beta N(0) \int_{-\omega}^{\omega} {\rm d}\xi \, \frac{1 + 2e^{-\beta\tilde{\xi}}}{4 + 3e^{-\beta\tilde{\xi}} + 2\cosh\beta\sqrt{\tilde{\xi}^2 + 2\Delta^2(T)}}, \tag{34}$$

where $\tilde{\xi} = \xi - \mu$. The temperature dependence of the spin susceptibility for the superconducting phase (34) is presented in Figure 11.

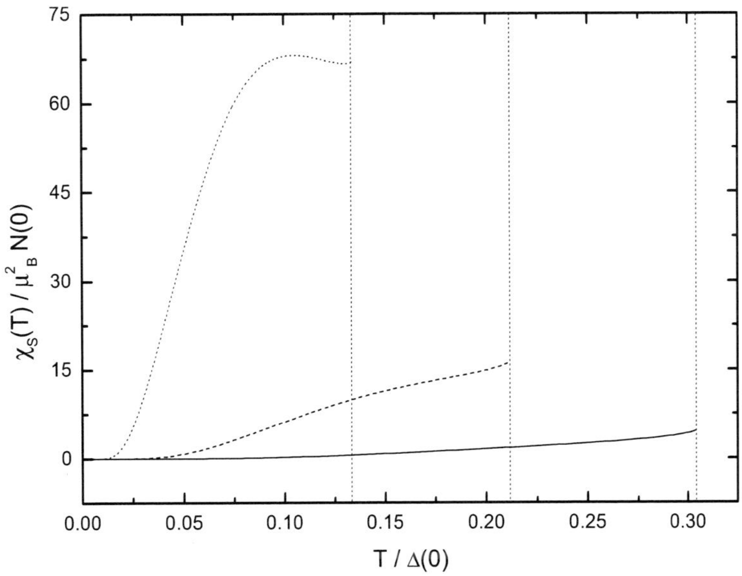

Figure 11. The superconducting-phase spin susceptibility of the singular Fermi liquid for $\omega/\Delta(0) = 2$ (solid), $2\sqrt{5}$ (dashed), and 10 (dotted). The dotted vertical lines mark the phase transition temperature $T_{\rm c}$ [11, 12].

For $\Delta_{\mathbf{k}} = 0$, *i.e.* for the normal phase, the magnetization and the spin susceptibility found from the partition function (33) read

$$\begin{aligned} M_{\rm N}(T,h) &= 2\mu_{\rm B}\sum_{\mathbf{k}} \frac{e^{-\beta\tilde{\epsilon}_{\mathbf{k}}}\sinh\beta h}{1 + 2e^{-\beta\tilde{\epsilon}_{\mathbf{k}}}\cosh\beta h}, \\ \chi_{\rm N}(T) &= \mu_{\rm B}^2\beta\sum_{\mathbf{k}} \frac{1}{1 + e^{\beta\tilde{\epsilon}_{\mathbf{k}} - \ln 2}}. \end{aligned} \tag{35}$$

Note that in the normal phase the total number of particles is [11, 41]

$$N = \sum_{\mathbf{k}} \frac{1}{1 + e^{\beta\tilde{\epsilon}_{\mathbf{k}} - \ln 2}}.$$

Hence the spin susceptibility of the normal phase $\chi_{\rm N}(T) = N\mu_{\rm B}^2/T$ diverges as $T \to 0$. Moreover at zero temperature the magnetization $M_{\rm N}(0,h) = \mu_{\rm B}\sum_{\mathbf{k}} 1 = 2N\mu_{\rm B}$, *i.e.* all spins are aligned along the field direction. These properties are analogous to those exhibited by the one-dimensional Ising model. According to the Van Hove scenario for the singular Fermi liquid with purely s-wave pairing and $N(\xi) \approx N(0)$ the formula (35) transforms into

$$\chi_{\rm N}(T) = \mu_{\rm B}^2 \beta N(0) \int_{-\omega}^{\omega} {\rm d}\xi \, \frac{1}{1 + e^{-\beta\tilde{\xi} - \ln 2}}. \tag{36}$$

The temperature dependence of the normal-phase spin susceptibility (36) is presented in Figure 12.

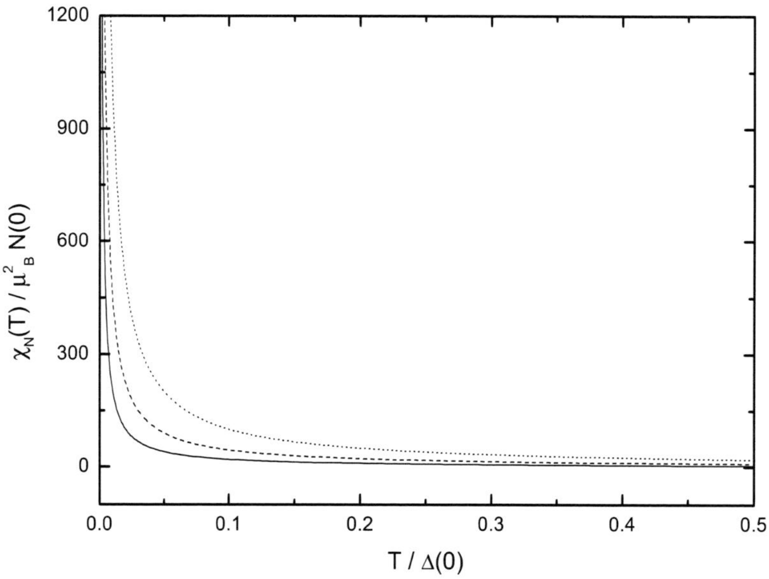

Figure 12. The normal-phase spin susceptibility of the singular Fermi liquid for $\omega/\Delta(0) = 2$ (solid), $2\sqrt{5}$ (dashed), and 10 (dotted) [11,12].

Note that the partition function (33) for the normal phase can be rewritten as

$$Z_{\rm N} = \prod_{\mathbf{k}} \left\{ 1 + \exp\left[-\beta \left(\tilde{\epsilon}_{\mathbf{k}} - T \ln 2 - T \ln \cosh \beta h \right) \right] \right\}.$$

Hence the chemical potential of the normal phase (measured from the Fermi level) in a magnetic field is $\mu_{\rm N}(T) = -T \ln 2 - T \ln \cosh \beta h$, and in the limit $H \to 0$ one has $\mu_{\rm N}(T) = -T \ln 2 - \mu_{\rm B}^2 H^2/2T$, *i.e.* the corrections due to the magnetic field are of the order of H^2. On the other hand in the limit $T \to 0$ the chemical potential $\mu_{\rm N}(0) = -\mu_{\rm B} H$, and the zero-temperature entropy

$$S_{\rm N}(0) = -N \left. \frac{\partial \mu(T)}{\partial T} \right|_{T=0}$$

is $N \ln 2$ for $H = 0$ and zero for $H > 0$, indicating a phase transition analogous to that in the one-dimensional Ising model.

8.2. London Penetration Depth

In order to describe the interaction of the magnetic field with charge degrees of freedom we introduce the superflow

$$\mathbf{v} = \nabla\chi - \frac{e}{m}\mathbf{A}, \tag{37}$$

where χ is the phase of the Cooper pairs wave function, $\mathbf{A}$ is the vector potential ($\mathbf{H} = \nabla \times \mathbf{A}$) and m can be treated either as the bare or the effective mass. The superflow is gauge-invariant and has a physical meaning of the Cooper pair velocity [14,42–45]. For the vector potential we choose the London gauge [46, 47], what implies that in the expansion of Gorkov equations the first correction to the energy gap vanishes. Consequently, the first correction to the Green function includes the energy gap derived in the absence of an external field [47]. The same applies to the linear response method [46, 48].

The inclusion of the superflow in the Hamiltonian is performed by a transformation of the particle momentum $\mathbf{k} \mapsto \mathbf{k} + m\mathbf{v}$, where $k \gg mv$. Hence, the total momentum of the Cooper pair is now $2m\mathbf{v}$, unlike in the absence of the magnetic field when it is zero. Consequently, the one-particle energy undergoes a transformation $\epsilon_{\mathbf{k}} \mapsto \epsilon_{\mathbf{k}+m\mathbf{v}}$. Moreover, the chemical potential $\mu \mapsto \mu - \frac{1}{2}mv^2$, *i.e.* is constant up to terms $O(v^2)$. The Hamiltonian taking into account the field-charge interaction by means of the superflow is therefore of the form

$$\begin{aligned} \hat{\kappa}_{\mathbf{k}+m\mathbf{v}} &= \tilde{\epsilon}_{\mathbf{k}+m\mathbf{v}}(\hat{n}_{\mathbf{k}+m\mathbf{v}\uparrow} + \hat{n}_{\mathbf{k}+m\mathbf{v}\downarrow}) + \tilde{\epsilon}_{-\mathbf{k}+m\mathbf{v}}(\hat{n}_{-\mathbf{k}+m\mathbf{v}\uparrow} + \hat{n}_{-\mathbf{k}+m\mathbf{v}\downarrow}) \\ &\quad - \Delta_{\mathbf{k}}\left(\hat{a}^{\dagger}_{\mathbf{k}+m\mathbf{v}\uparrow}\hat{a}^{\dagger}_{-\mathbf{k}+m\mathbf{v}\downarrow} + \hat{a}^{\dagger}_{-\mathbf{k}+m\mathbf{v}\uparrow}\hat{a}^{\dagger}_{\mathbf{k}+m\mathbf{v}\downarrow} + \text{H.c.}\right). \end{aligned} \tag{38}$$

Let us note that now—unlike in the previous subsection—one has to define a new basis for the 9-dimensional Fock space, namely

$$\psi_{\sigma\,\sigma'}(\mathbf{k}) = \left| \begin{array}{cc} \mathbf{k}+m\mathbf{v}, & \sigma \\ -\mathbf{k}+m\mathbf{v}, & \sigma' \end{array} \right\rangle, \tag{39}$$

where $\sigma, \sigma' = 0, \uparrow$ or $\downarrow$. Matrix elements of the Hamiltonian (38) in the basis (39) are given in Table 9.

Table 9. Matrix elements of the Hamiltonian (38); off-diagonal zeros omitted [11].

	$0,0$	$\uparrow,0$	$\downarrow,0$	$0,\uparrow$	$\uparrow,\uparrow$	$\downarrow,\uparrow$	$0,\downarrow$	$\uparrow,\downarrow$	$\downarrow,\downarrow$
$0,0$	0					$-\Delta_{\mathbf{k}}$		$-\Delta_{\mathbf{k}}$	
$\uparrow,0$		$\tilde{\epsilon}_{\mathbf{k}}+\mathbf{kv}$							
$\downarrow,0$			$\tilde{\epsilon}_{\mathbf{k}}+\mathbf{kv}$						
$0,\uparrow$				$\tilde{\epsilon}_{\mathbf{k}}-\mathbf{kv}$					
$\uparrow,\uparrow$					$2\tilde{\epsilon}_{\mathbf{k}}$				
$\downarrow,\uparrow$	$-\Delta_{\mathbf{k}}$					$2\tilde{\epsilon}_{\mathbf{k}}$			
$0,\downarrow$							$\tilde{\epsilon}_{\mathbf{k}}-\mathbf{kv}$		
$\uparrow,\downarrow$	$-\Delta_{\mathbf{k}}$							$2\tilde{\epsilon}_{\mathbf{k}}$	
$\downarrow,\downarrow$									$2\tilde{\epsilon}_{\mathbf{k}}$

Table 10. Eigenvalues and eigenstates of the Hamiltonian (38) [11].

eigenvalue	*eigenstate*
$\tilde{\epsilon}_{\mathbf{k}} - E_{\mathbf{k}}$	$\sqrt{\frac{E_{\mathbf{k}}+\tilde{\epsilon}_{\mathbf{k}}}{2E_{\mathbf{k}}}}\,\psi_{00}(\mathbf{k}) - \frac{\sqrt{2}}{2}\sqrt{\frac{E_{\mathbf{k}}-\tilde{\epsilon}_{\mathbf{k}}}{2E_{\mathbf{k}}}}\,[\psi_{\downarrow\uparrow}(\mathbf{k}) + \psi_{\uparrow\downarrow}(\mathbf{k})]$
$\tilde{\epsilon}_{\mathbf{k}} + \mathbf{kv}$	$\psi_{\uparrow 0}(\mathbf{k})$
$\tilde{\epsilon}_{\mathbf{k}} + \mathbf{kv}$	$\psi_{\downarrow 0}(\mathbf{k})$
$\tilde{\epsilon}_{\mathbf{k}} - \mathbf{kv}$	$\psi_{0\uparrow}(\mathbf{k})$
$2\tilde{\epsilon}_{\mathbf{k}}$	$\psi_{\uparrow\uparrow}(\mathbf{k})$
$\tilde{\epsilon}_{\mathbf{k}} + E_{\mathbf{k}}$	$\sqrt{\frac{E_{\mathbf{k}}-\tilde{\epsilon}_{\mathbf{k}}}{2E_{\mathbf{k}}}}\,\psi_{00}(\mathbf{k}) + \frac{\sqrt{2}}{2}\sqrt{\frac{E_{\mathbf{k}}+\tilde{\epsilon}_{\mathbf{k}}}{2E_{\mathbf{k}}}}\,[\psi_{\downarrow\uparrow}(\mathbf{k}) + \psi_{\uparrow\downarrow}(\mathbf{k})]$
$\tilde{\epsilon}_{\mathbf{k}} - \mathbf{kv}$	$\psi_{0\downarrow}(\mathbf{k})$
$2\tilde{\epsilon}_{\mathbf{k}}$	$\frac{\sqrt{2}}{2}\,[\psi_{\downarrow\uparrow}(\mathbf{k}) - \psi_{\uparrow\downarrow}(\mathbf{k})]$
$2\tilde{\epsilon}_{\mathbf{k}}$	$\psi_{\downarrow\downarrow}(\mathbf{k})$

In evaluation of the matrix elements it has been employed that $\epsilon_{\mathbf{k}+m\mathbf{v}} \approx \epsilon_{\mathbf{k}} + m\mathbf{v}\frac{\partial \epsilon_{\mathbf{k}}}{\partial \mathbf{k}}$. Moreover, including that $m\nabla_{\mathbf{k}}\epsilon_{\mathbf{k}} = \mathbf{k}$, one has $\epsilon_{\mathbf{k}+m\mathbf{v}} \approx \epsilon_{\mathbf{k}} + \mathbf{kv}$ and $\epsilon_{-\mathbf{k}+m\mathbf{v}} \approx \epsilon_{\mathbf{k}} - \mathbf{kv}$.

The diagonalization of the Hamiltonian yields eigenvalues and eigenstates (*cf.* Table 10), and allows us to find the partition function

$$Z = \prod_{\mathbf{k}} \left\{ \left[3e^{-\beta\tilde{\epsilon}_{\mathbf{k}}} + 4\cosh\beta\mathbf{kv} + 2\cosh\beta E_{\mathbf{k}} \right] e^{-\beta\tilde{\epsilon}_{\mathbf{k}}} \right\}^{\frac{1}{2}}. \tag{40}$$

In the previous section the energy of the Pauli interaction has been explicitly included in the Hamiltonian (31) in the form of one-particle interaction. However, in the present case we have to include the energy of the magnetic field itself in the thermodynamic potential. The variation of the thermodynamic potential due to the magnetic field is $\delta\Omega = \Omega' - \Omega = \frac{1}{4\pi}\mathbf{H}\cdot\delta\mathbf{H}$, where Ω' denotes the thermodynamic potential in the presence of the magnetic field. Employing the relation between the magnetic field and the vector potential as well as the static Maxwell equation $\nabla \times \mathbf{H} = 4\pi\mathbf{j}$, we eventually obtain

$$\delta\Omega = \mathbf{j}\cdot\delta\mathbf{A}, \tag{41}$$

where $\mathbf{j}$ is the current of normal particles, which can be derived from the partition function by employing Eqs. (8), (37) and (41)

$$\mathbf{j} = \frac{e}{m\beta}\frac{\partial \ln Z}{\partial \mathbf{v}}. \tag{42}$$

The superconducting current of Cooper pairs is given by

$$\mathbf{j}_{\mathrm{S}} = eN\mathbf{v} - \mathbf{j}. \tag{43}$$

The partition function (40) and Eq. (42) allow us to calculate the superconducting current in the singular Fermi liquid as

$$\mathbf{j}_{\mathrm{S}} = eN\mathbf{v} - \frac{e\beta}{2m}\sum_{\mathbf{k}} \mathbf{k}\frac{\sinh\beta\mathbf{kv}}{\cosh\beta\mathbf{kv} + \frac{3}{4}e^{-\beta\tilde{\epsilon}_{\mathbf{k}}} + \frac{1}{2}\cosh\beta E_{\mathbf{k}}}.$$

Expanding this expression to the first order in $\mathbf{v}$ we obtain

$$\mathbf{j}_\mathrm{S} = eN\mathbf{v} - \frac{e\beta}{2m}\sum_{\mathbf{k}} \mathbf{k}(\mathbf{k}\mathbf{v})\frac{4}{4+3e^{-\beta\tilde{\epsilon}_\mathbf{k}}+2\cosh\beta E_\mathbf{k}},$$

where the next term is $O(\mathbf{v}^3)$. For the singular Fermi liquid with purely s-wave pairing and $N(\xi) \approx N(0)$ we eventually obtain

$$j_\mathrm{S} = eNv\left[1-\beta\gamma\int_{-\omega}^{\omega}\frac{\mathrm{d}\xi}{4+3e^{-\beta\tilde{\xi}}+2\cosh\beta\sqrt{\tilde{\xi}^2+2\Delta^2(T)}}\right], \tag{44}$$

where also the relations $N = N(0)\omega$ and $\gamma \simeq m|\nabla\epsilon_\mathbf{k}|^2/\omega \approx 1$ have been taken into account [49]. Note that Eq. (43) can be rewritten as

$$\mathbf{j}_\mathrm{S} = eN\mathbf{v}\left(1-\frac{N_\mathrm{N}}{N}\right),$$

where N_N is the number of normal particles. Hence the latter term in Eq. (44) determines the temperature dependent fraction $N_\mathrm{N}(T)/N$ of normal particles. For $T = 0$, when all particles form Cooper pairs, we have $N_\mathrm{N}(0) = 0$. Moreover, this quantity is related with the London penetration depth of the magnetic field by the formula [46]

$$\left[\frac{\lambda(0)}{\lambda(T)}\right]^2 = 1-\frac{N_\mathrm{N}(T)}{N}.$$

Hence for a purely s-wave paired singular Fermi liquid the London penetration depth is given by

$$\left[\frac{\lambda(0)}{\lambda(T)}\right]^2 = 1-\beta\int_{-\omega}^{\omega}\frac{\mathrm{d}\xi}{4+3e^{-\beta\tilde{\xi}}+2\cosh\beta\sqrt{\tilde{\xi}^2+2\Delta^2(T)}}.$$

The temperature dependence of the London penetration depth is presented in Figure 13.

8.3. Critical Magnetic Field

Having known the values of the free energy for both phases in the absence of any external fields (*cf.* Section 5.) one can easily find the critical magnetic field H_c for any value of the temperature [14, 46]

$$H_\mathrm{c}(T) = \sqrt{\frac{F_\mathrm{N}(T)-F_\mathrm{S}(T)}{8\pi}}. \tag{45}$$

The temperature dependence of the critical magnetic field is presented in Figure 14.

Rewriting Eq. (45) as $F_\mathrm{N}(T) - F_\mathrm{S}(T) = H_\mathrm{c}^2(T)/8\pi$ and differentiating both sides with respect to the temperature we obtain

$$S_\mathrm{S}(T) - S_\mathrm{N}(T) = \frac{1}{4\pi}H_\mathrm{c}(T)\frac{\mathrm{d}H_\mathrm{c}(T)}{\mathrm{d}T}. \tag{46}$$

The rhs of Eq. (46) is negative for all $T \neq T_\mathrm{c}$ (*cf.* Figure 14), hence the superconducting phase is the phase of the lower entropy. Moreover, Eq. (46) allows us to find the heat of the transition, which is non-zero for in the discussed temperature range. Thus, the phase transition between the superconducting and the normal phase driven by the magnetic field is of the first order like in the case $H = 0$, at $T = T_\mathrm{c}$.

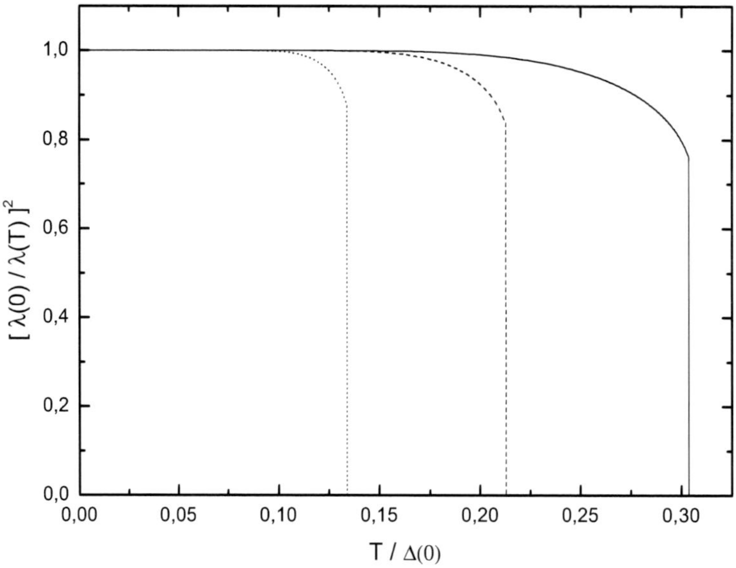

Figure 13. The London penetration depth for $\omega/\Delta(0) = 2$ (solid), $2\sqrt{5}$ (dashed), and 10 (dotted) in the singular Fermi liquid model [11].

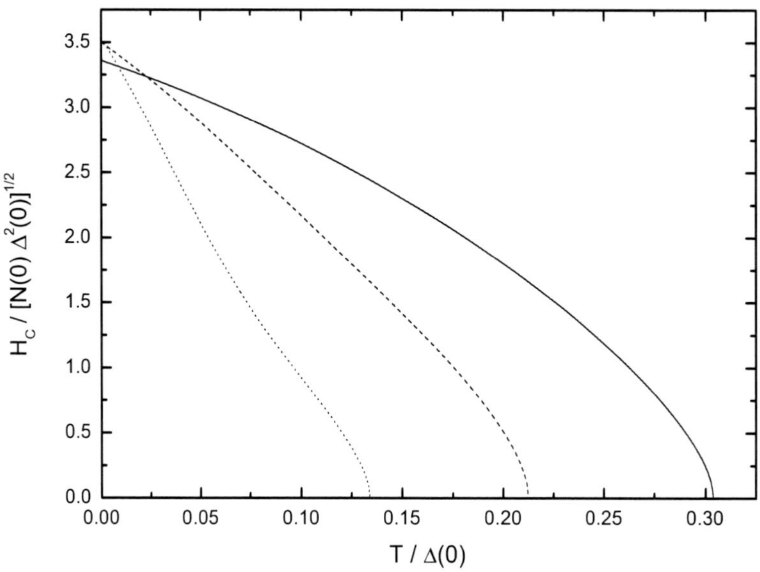

Figure 14. The critical magnetic field for $\omega/\Delta(0) = 2$ (solid), $2\sqrt{5}$ (dashed), and 10 (dotted) in the singular Fermi liquids model [11, 13].

8.4. Critical Current

In the limit $k \gg mv$ we can expand the free energy into a series, conserving only the terms up to $O(v^2)$. We then obtain the following formula for the free energy $F(\mathbf{v}) = \Omega(\mathbf{v}) + \mu N$

in the presence of the superflow (37)

$$F(\mathbf{v}) = \Omega(0) + \sum_{i=1}^{3} \left.\frac{\delta\Omega(\mathbf{v})}{\delta v_i}\right|_{\mathbf{v}=0} v_i + \frac{1}{2}\sum_{i,j=1}^{3} \left.\frac{\delta^2\Omega(\mathbf{v})}{\delta v_j \delta v_j}\right|_{\mathbf{v}=0} v_j v_i + \mu N, \tag{47}$$

where $\delta/\delta v_i$ denotes the variational derivative. Eqs. (37) and (41) imply

$$\frac{\delta\Omega(\mathbf{v})}{\delta v_i} = -\frac{m}{e} j_i = -mN_{\mathrm{N}} v_i$$

and hence

$$\left.\frac{\delta\Omega(\mathbf{v})}{\delta v_i}\right|_{\mathbf{v}=0} = 0 \quad \text{and} \quad \left.\frac{\delta^2\Omega(\mathbf{v})}{\delta v_i \delta v_j}\right|_{\mathbf{v}=0} = \left.\frac{\delta}{\delta v_i}(-mN_{\mathrm{N}} v_j)\right|_{\mathbf{v}=0} = -mN_{\mathrm{N}}\delta_{ij}.$$

Employing the calculated variational derivatives we can rewrite Eq. (47) as

$$F(\mathbf{v}) = \Omega(0) - \frac{m}{2e}\mathbf{j}\cdot\mathbf{v} + \mu N.$$

Since the coherent state of Cooper pairs does not exist in the normal phase, the electric current is given by $\mathbf{j} = eN\mathbf{v}$, whereas for the superconducting phase it is $\mathbf{j} = eN\mathbf{v} - \mathbf{j}_{\mathrm{S}}$ (*cf.* Eq. (43)). Hence the free energy for both phases in the presence of the superflow can be approximated by

$$\begin{aligned} F_{\mathrm{N}}(\mathbf{v}) &= \Omega_{\mathrm{N}}(0) - \frac{1}{2}mN\mathbf{v}^2 + \mu_{\mathrm{N}} N, \\ F_{\mathrm{S}}(\mathbf{v}) &= \Omega_{\mathrm{S}}(0) - \frac{1}{2}mN\mathbf{v}^2 + \frac{m}{2e}\mathbf{j}_{\mathrm{S}}\cdot\mathbf{v} + \mu_{\mathrm{S}} N. \end{aligned}$$

The free energy difference $\Delta F(\mathbf{v}) = F_{\mathrm{S}}(\mathbf{v}) - F_{\mathrm{N}}(\mathbf{v})$ in the presence of the superflow is hence

$$\Delta F(\mathbf{v}) = \Delta F(0) + \frac{m}{2e}\mathbf{j}_{\mathrm{S}}\cdot\mathbf{v} = \Delta F(0) + \frac{1}{2}m(N - N_{\mathrm{N}})v^2, \tag{48}$$

where $\Delta F(0)$ is the corresponding difference in the absence of the superflow.

The persistent current of Cooper pairs $\mathbf{j}_{\mathrm{s}}$ flowing through the system in the superconducting phase destroys superconductivity if the superflow reaches the critical value $\mathbf{v}_{\mathrm{c}}$, for which $\Delta F(\mathbf{v}_{\mathrm{c}}) = 0$. This condition along with Eq. (48) allows us to find the critical superflow value

$$v_{\mathrm{c}} = \sqrt{-\frac{2\Delta F(0)}{m(N - N_{\mathrm{N}})}}.$$

The corresponding Cooper pairs current (the so-called critical current) $\mathbf{j}_{\mathrm{c}} = e(N - N_{\mathrm{N}})\mathbf{v}_{\mathrm{c}}$, at which the system undergoes a phase transition to the normal phase, is given by

$$j_{\mathrm{c}} = e\sqrt{-\frac{2\Delta F(0)(N - N_{\mathrm{N}})}{m}}. \tag{49}$$

Since for $T = 0$ all particles are paired ($N_{\mathrm{n}} = 0$), the chemical potentials $\mu_{\mathrm{S}}(0) = \mu_{\mathrm{N}}(0) = 0$, and $\Delta F(0) = -N(0)\Delta^2(0)/2$, for $\omega \gg \Delta(0)$ [9], Eq. (49) simplifies to the form $j_{\mathrm{c}} = e\sqrt{N\,N(0)/m}\Delta(0)$.

Employing the calculations of the normal particles fraction $N_{\mathrm{N}}(T)/N$ from Section 8.2. we obtain the formula for the critical current in the singular Fermi liquid system with purely s-wave pairing

$$j_{\mathrm{c}} = e\sqrt{-\frac{2\Delta F(0)N}{m}\left(1-\beta\int\limits_{-\omega}^{\omega}\frac{\mathrm{d}\xi}{4+3e^{-\beta\tilde{\xi}}+2\cosh\beta\sqrt{\tilde{\xi}^2+2\Delta^2(T)}}\right)}.$$

The temperature dependence of the critical current is presented in Figure 15.

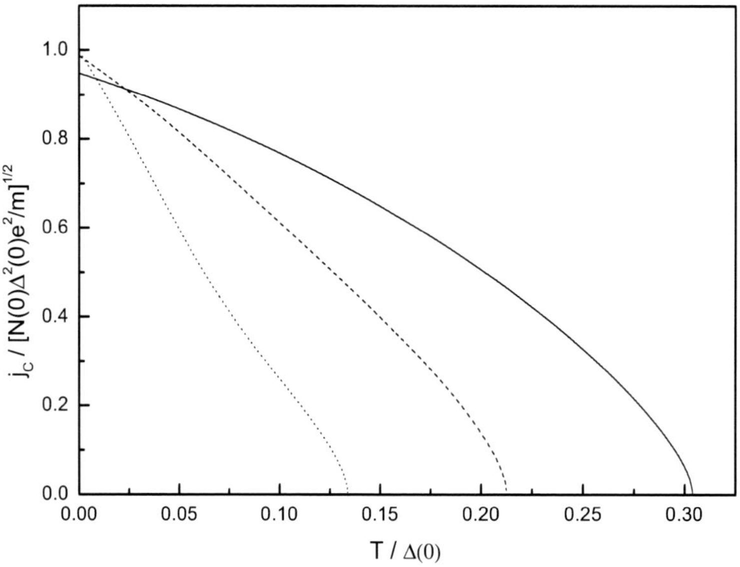

Figure 15. The critical current for $\omega/\Delta(0) = 2$ (solid), $2\sqrt{5}$ (dashed), and 10 (dotted) in the singular Fermi liquid model [11, 13].

9. Conclusion

In this chapter we considered a model of a singular Fermi liquid with divergent amplitude for particles (electrons or holes) with the same momenta [9–13]. In the model a pseudogap-type behavior can be observed with two characteristic temperatures: T^* — the temperature at which the gap in the excitation spectrum opens, but the system remains in the normal phase, and the temperature T_{c} (lower than T^*), at which the free energy of the superconducting phase and the free energy of the normal phase become equal, the Cooper pairs gain the phase coherence and the system turns into the stable superconducting phase (*cf.* Table 5). The phase transition between the superconducting and the normal phase is of the first order (*cf.* Figure 7 and Table 6). It is the result of the competition between the free energy of the normal phase (with a large residual entropy $S_{\mathrm{N}}(0) = N\ln 2$) with the entropy increasing with the temperature and—on the other hand—the free free energy of the superconducting phase (without the residual entropy), which at $T = 0$ is lowered due to the non-zero order parameter. For the system at $T = T_c$ it is energetically more advantageous to posses a large entropy than to keep a non-zero order parameter. Note that this effect is analogous to the first-order phase transition driven by the magnetic field in type-I superconductors, where the strong magnetic field can decrease the free energy of the system more than a non-zero energy gap in the superconducting phase. Such effect cannot be observed in systems with the entropy of the normal phase vanishing at $T = 0$. In this case the phase transition has to be of the second order.

Including the external magnetic field by means of the Pauli term allowed us to evaluate its interaction with the system's spin degrees of freedom. The magnetic susceptibility of the superconducting phase characterizing the response of the system is especially pronounced for a wide band (*cf.* Figure 11). On the other hand, the states in the normal phase are being occupied singly and the phase is paramagnetic. Applying an external magnetic field results in the total (if $T = 0$) or partial (if $T > 0$) alignment of spins. Moreover, the normal-phase spin susceptibility has a Curie-law character (*cf.* Figure 12).

The study of the interaction of the applied field with charge degrees of freedom has proved that there is only a small fraction of normal particles close to the phase transition temperature $T = T_c$ (*cf.* Figure 13). Consequently, the magnetic field penetration depth does not achieve large values, unlike in the BCS case when it tends to infinity at $T = T_c$. Since the penetration depth characterizes the region of the superconductor in which screening currents flow—preventing the magnetic field from entering into the superconductor—one can state that this region is rather narrow in the studied system.

The differences of the zero-temperature values of the critical magnetic field (*cf.* Figure 14) and the critical current (*cf.* Figure 15) for various values of the band width are due to the dependence of the free energy difference for both—the superconducting and the normal—phases at $T = 0$ on the parameter ω (*cf.* Eq. (28) and Figure 1).

It is worth to emphasize that employing the developed formalism it is possible to study more complex systems, *e.g.* with anisotropic pairing interaction (anisotropic order parameter) and an arbitrary dispersion relation.

Acknowledgements

One of the authors (MK) is grateful to Professor Shiping Feng for his kind hospitality at the Department of Physics, Beijing Normal University. MK was supported by the China Scholarship Council (CSC No. 2007616032). Research scholarship from Wroclaw University of Technology is also gratefully acknowledged.

A Landau–Pomeranchuk Instabilities

The interaction in the particle-hole channel, including the Fermi-liquid corrections for two- and three-dimensional systems, can be respectively taken into account in the form

$$G^{\mathrm{s,a}}(\hat{\mathbf{k}}\cdot\hat{\mathbf{k}}') = G_0^{\mathrm{s,a}} + 2\sum_{l=1}^{\infty} G_l^{\mathrm{s,a}} \cos l(\varphi - \varphi'), \tag{50}$$

$$F^{\mathrm{s,a}}(\hat{\mathbf{k}}\cdot\hat{\mathbf{k}}') = \sum_{l=0}^{\infty} F_l^{\mathrm{s,a}} P_l(\hat{\mathbf{k}}\cdot\hat{\mathbf{k}}'), \tag{51}$$

where $\hat{\mathbf{k}} \cdot \hat{\mathbf{k}}' = \cos(\varphi - \varphi')$, P_l denotes the Legendre polynomial, $G_l^{\mathrm{s,a}}$ and $F_l^{\mathrm{s,a}}$ are the Landau parameters satisfying the relations [4–7, 50]

$$-1 < G_l^{\mathrm{s,a}} < \infty \qquad \text{or} \qquad -(2l+1) < F_l^{\mathrm{s,a}} < \infty.$$

Moreover, in order to ensure the convergence of the interactions, the parameters $G_l^{\mathrm{s,a}}$ and $F_l^{\mathrm{s,a}}$ should vanish sufficiently fast as $l \to \infty$. Introducing the scattering amplitudes [7]

$$\begin{aligned} R^{\mathrm{s,a}}(\hat{\mathbf{k}} \cdot \hat{\mathbf{k}}') &= r_0^{\mathrm{s,a}} + 2\sum_{l=1}^{\infty} r_l^{\mathrm{s,a}} \cos l(\varphi - \varphi'), \\ R^{\mathrm{s,a}}(\hat{\mathbf{k}} \cdot \hat{\mathbf{k}}') &= \sum_{l=0}^{\infty} R_l^{\mathrm{s,a}} P_l(\hat{\mathbf{k}} \cdot \hat{\mathbf{k}}'), \end{aligned}$$

for two- and three-dimensional systems, respectively, we obtain two functions playing the same role as the functions (50) and (51).

The relations between the terms in Eqs. (50), (52) and Eqs. (51), (52)

$$r_l^{\mathrm{s,a}} = \frac{G_l^{\mathrm{s,a}}}{1 + G_l^{\mathrm{s,a}}}, \qquad R_l^{\mathrm{s,a}} = \frac{F_l^{\mathrm{s,a}}}{1 + \frac{1}{2l+1} F_l^{\mathrm{s,a}}},$$

imply the stability conditions of the form

$$-\infty < r_l^{\mathrm{s,a}} < 1 \qquad \text{and} \qquad -\infty < R_l^{\mathrm{s,a}} < 2l+1.$$

Because $G_l^{\mathrm{s,a}}, F_l^{\mathrm{s,a}} \to 0$ for $l \to \infty$, also the coefficients $r_l^{\mathrm{s,a}}$, $R_l^{\mathrm{s,a}}$ tend to zero for large values of l. Hence, if the above relations hold, the regular Fermi liquid is stable in the considered quantum system.

On the other hand if, *e.g.* in a two-dimensional system, $r_l^{\mathrm{s}} \to r_\infty^{\mathrm{s}} > 0$ for $l \to \infty$, the the scattering amplitude

$$\begin{aligned} R^{\mathrm{s}}(\hat{\mathbf{k}} \cdot \hat{\mathbf{k}}') &= r_0^{\mathrm{s}} - r_\infty^{\mathrm{s}} + 2\sum_{l=1}^{\infty} (r_l^{\mathrm{s}} - r_\infty^{\mathrm{s}}) \cos l(\varphi - \varphi') + r_\infty^{\mathrm{s}} \sum_{l=-\infty}^{\infty} e^{il(\varphi - \varphi')} \\ &= r_0^{\mathrm{s}} - r_\infty^{\mathrm{s}} + 2\sum_{l=1}^{\infty} (r_l^{\mathrm{s}} - r_\infty^{\mathrm{s}}) \cos l(\varphi - \varphi') + r_\infty^{\mathrm{s}} \delta(\varphi - \varphi'). \end{aligned}$$

Hence

$$R_{\mathbf{k}} \delta_{\mathbf{k}\mathbf{k}'} \sim r_\infty \delta(\mathbf{k} - \mathbf{k}'),$$

and this instability corresponds to the singular Fermi liquid.

The generating function for the Legendre polynomials is of the form

$$\sum_{l=0}^{\infty} t^l P_l(\hat{\mathbf{k}} \cdot \hat{\mathbf{k}}') = \frac{1}{\sqrt{1 - 2(\hat{\mathbf{k}} \cdot \hat{\mathbf{k}}')t + t^2}},$$

where $0 \le t < 1$ and $|\ \hat{\mathbf{k}} \cdot \hat{\mathbf{k}}'\ | \le 1$. In the limit $t \to 1$ the series is divergent if $|\ \hat{\mathbf{k}} - \hat{\mathbf{k}}'\ | \to 0$. This is also valid for the three-dimensional case, if $R_l^{\mathrm{s}} \to R_\infty^{\mathrm{s}} > 0$ for $l \to \infty$. Hence, the instabilities of the Landau–Pomeranchuk type can lead to various form of non-Fermi liquids.

B Dominating Types of Binary Collisions in the Vicinity of the Fermi Surface

For particles with a parabolic (linear) energy spectrum participating in binary collisions on the Fermi surface (line) the following relations are fulfilled

$$|\mathbf{p}| = |\mathbf{p}' + \mathbf{q}| = |\mathbf{p}'| = |\mathbf{p} + \mathbf{q}| = p_{\mathrm{F}}. \tag{52}$$

Representing the vector $\mathbf{q}$ as $\mathbf{q} = a\mathbf{p} + b\mathbf{p}'$, where a and b are real numbers, from Eq. (52) we obtain a system of two equations

$$\begin{aligned} (1+a)^2 + b^2 + 2(1+a)b\cos\theta &= 1, \\ (1+b)^2 + a^2 + 2(1+b)a\cos\theta &= 1, \end{aligned}$$

where θ is the angle between the vectors $\mathbf{p}$ and $\mathbf{p}'$. These equations can be also written as

$$\begin{aligned} (1+a)a + (1+b)b + [(1+a)b + (1+b)a]\cos\theta &= 0, \\ (a-b)(1-\cos\theta) &= 0. \end{aligned}$$

For an arbitrary value of θ there exist two independent solutions

$$a = b = 0 \qquad \text{and} \qquad a = b = -1.$$

corresponding to $\mathbf{q} = 0$ (*i.e.* $\mathbf{p} + \mathbf{q} = \mathbf{p}$ and $\mathbf{p}' + \mathbf{q} = \mathbf{p}'$), and $\mathbf{q} = -\mathbf{p} - \mathbf{p}'$ (*i.e.* $\mathbf{p} + \mathbf{q} = -\mathbf{p}'$, $\mathbf{p}' + \mathbf{q} = -\mathbf{p}$), respectively. The former solution refers to the case of the momentum transfer equal to zero, whereas the latter one to the interaction of particles with opposite momenta.

Further consideration in the case when the vectors $\mathbf{p}$ and $\mathbf{p}'$ lie in the vicinity of the Fermi surface (line), *i.e.*

$$\begin{aligned} |\mathbf{p}| &= (1+c_1)p_{\mathrm{F}}, \qquad & |\mathbf{p}' + \mathbf{q}| &= (1+c_2)p_{\mathrm{F}}, \\ |\mathbf{p}'| &= (1+c_3)p_{\mathrm{F}}, \qquad & |\mathbf{p} + \mathbf{q}| &= (1+c_4)p_{\mathrm{F}}, \end{aligned}$$

when $|c_i| \ll 1$, $i = 1, 2, 3, 4$, should be performed within the perturbation method assuming that

$$\mathbf{q} = (a + a')\mathbf{p} + (b + b')\mathbf{p}',$$

where a' and b' are of the same order of magnitude as c_i, *i.e.* $|a'|, |b'| \ll 1$. Applying the previous method and the energy conservation law we arrive at the system of equations

$$\begin{aligned} (1+a)a' + b\,b' + (1+a)^2 c_1 + b^2 c_3 &= c_4, \\ (1+a)b' + b\,a' + (1+a)\,b\,(c_1 + c_3) &= 0, \\ (1+b)b' + a\,a' + (1+b)^2 c_3 + a^2 c_1 &= c_2, \\ (1+b)a' + a\,b' + (1+b)\,a\,(c_1 + c_3) &= 0, \end{aligned}$$

with

$$c_1 + c_2 = c_3 + c_4.$$

These equations, after taking into account that $a = b = 0$, simplify to the form

$$a' = b' = 0 \qquad \text{and} \qquad c_1 = c_4, \qquad c_2 = c_3.$$

Hence $\mathbf{q} = 0$, although in general $|\mathbf{p}| \neq |\mathbf{p}'|$. On the other hand for $a = b = -1$ we obtain

$$a' = b' = 0 \qquad \text{and} \qquad c_1 = c_2 = c_3 = c_4.$$

Hence $\mathbf{q} = -\mathbf{p} - \mathbf{p}'$ and $|\mathbf{p}| = |\mathbf{p}'|$, so that $\xi_{\mathbf{p}} = \xi_{\mathbf{p}'}$.

The presented results prove that solely two types of interaction are favored and privileged in the system with a parabolic dispersion relation. The first one is in the spirit of the Fermi liquid theory, and the other one is responsible for creation of Cooper pairs of a definite symmetry if at least one term of the series expansion of the interaction potential is negative. On the other hand after performing the reverse conformal transformation we note again that the collisions with the momentum transfer equal to zero and the collisions involving particles with opposite momenta dominate. Hence, because Eq. (52) implies

$$\epsilon_{\mathbf{k}} = \epsilon_{\mathbf{k}'+\mathbf{q}'} = \epsilon_{\mathbf{k}'} = \epsilon_{\mathbf{k}+\mathbf{q}'} = \epsilon_{\mathrm{F}},$$

we can expect that for complex forms of the dispersion relation, arbitrary vectors $\mathbf{k}$,$\mathbf{k}'$ and all possible vectors $\mathbf{q}'$ this equality holds if $\mathbf{q}' = 0$ or $\mathbf{q}' = -\mathbf{k} - \mathbf{k}'$.

References

[1] Anderson, P. W. *Science* 2000, 288, 480.

[2] Pines, D. *Physica C* 2000, 341-348, 59.

[3] Manske, D.; Bennemann, K. H. *Physica C* 2000, 341-348, 83.

[4] Varma, C. M.; Nussinov, Z.; van Saarloos, W. *Phys. Rep.* 2002, 361, 267.

[5] Pomeranchuk, I. Y. *Zh. Eksp. Teor. Fiz.* 1958, 34, 524.

[6] Gonczarek, R.; Krzyzanowski, I. H., *Phys. Lett. A* 1992 171, 107.

[7] Platzman, P. M.; Wolff, P. A. *Waves and Interactions in Solid State Plasmas*, Academic Press, 1973.

[8] Misochko, O. V.; Georgiev, N.; Dekorsy, T.; Helm, M. *Phys. Rev. Lett.* 2002, 89, 067002.

[9] Gonczarek, R.; Krzyzosiak, M. *Int. J. Mod. Phys. B* 2003, 17, 5683.

[10] Gonczarek, R.; Krzyzosiak, M. *phys. stat. sol. (b)* 2004, 241, 352.

[11] Krzyzosiak, M.; Gonczarek, R. *Int. J. Mod. Phys. B* 2004, 18, 4143.

[12] Krzyzosiak, M.; Gonczarek, R. *Physica C* 2005, 426-431, 273.

[13] Gonczarek, R.; Krzyzosiak, M. *Physica C* 2006, 445-448, 158.

[14] Micnas, R.; Ranniger, J.; Robaszkiewicz, S. *Rev. Mod. Phys.* 1990, 62, 113.

[15] Spałek, J.; Wójcik, W. *Phys. Rev. B* 1988, 37, 1532.

[16] Spałek, J. *Physica B* 1990, 163, 621.

[17] Byczuk, K.; Spałek, J.; Wójcik, W. *Physica C* 1994, 235-240, 2347.

[18] Byczuk, K.; Spałek, J. *Phys. Rev. B* 1994, 50, 11403.

[19] Spałek, J. *Acta Phys. Pol. A* 1994, 85, 39.

[20] Spałek, J.; Wójcik, W. *Acta Phys. Pol. A* 1994, 85, 357.

[21] Byczuk, K.; Spałek, J. *Acta Phys. Pol. A* 1994, 85, 337.

[22] Byczuk, K.; Spałek, J. *Phys. Rev. B* 1995, 51, 7934.

[23] Karbowski, J.; Spałek, J. *Acta Phys. Pol. A* 1994, 85, 341.

[24] Joyce, G. S.; Sarkar, S.; Spałek, J.; Byczuk, K. *Phys. Rev. B* 1996, 53, 990.

[25] Hyyppa, J.; Byczuk, K.; Spałek, J.; Honig, J. M. *Solid State Comm.* 1996, 98, 367.

[26] Byczuk, K.; Spałek, J.; Wójcik, W. *Molecular Phys. Rep.* 1995, 12, 219.

[27] Byczuk, K.; Spałek, J.; Wójcik, W. *Acta Phys. Pol. B* 1998, 29, 3871.

[28] Gonczarek, R.; Krzyzosiak, M. *Physica C* 2005, 426-431, 278.

[29] Gonczarek, R.; Jacak, L.; Krzyzosiak, M.; Gonczarek, A. *Europ. Phys. J. B* 2006, 49, 171.

[30] Gonczarek, R.; Jacak, L.; Krzyzosiak, M.; Gonczarek, A. *Acta Phys. Pol. A* 2006, 109, 493.

[31] Anderson, P. W. *Phys. Rev. Lett.* 1990, 65, 2306.

[32] Engelbrecht, J. R.; Randeria, M.; Zhang L. *Phys. Rev. B* 1992, 45, 10135.

[33] Engelbrecht, J. R.; Randeria, M. *Phys. Rev. B* 1992 45, 12419.

[34] Bogolubov (Jr.), N. N. *Method for Studying Model Hamiltonians*, Nauka, Moscow, 1974, [in Russian].

[35] Bogolubov (Jr.), N. N.; Sadovnikov, B. I. *Selected problems in statistical mechanics*, Vyzhshaya Shkola, Moscow, 1975, [in Russian].

[36] Czerwonko, J. *Physica C* 1994, 235-240, 2337.

[37] Gonczarek, R.; Krzyzosiak, M. *phys. stat. sol. (b)* 2003, 238, 29.

[38] Gonczarek, R.; Krzyzosiak, M.; Mulak, M. *J. Phys. A: Math. Gen.* 2004, 37, 4899.

[39] Dorbolo, S.; Houssa, M.; Ausloos, M. *Physica C* 1996, 267, 24.

[40] Gonczarek, R.; Mulak, M. *phys. stat. sol. (b)* 1998, 208, 457.

[41] Gonczarek, R.; Gładysiewicz, M.; Mulak, M. *Int. J. Mod. Phys. B* 2001, 15, 491.

[42] Micnas, R.; Pedersen, M. H.; Schafroth, S.; Schneider, T.; Rodriguez-Nunez, J. J.; Beck, H. *Phys. Rev. B* 1995, 52, 16223.

[43] Micnas, R.; Pedersen, M. H.; Schafroth, S.; Schneider, T.; Rodriguez-Nunez, J. J.; Beck, H. *Phys. Rev. B* 1996, 54, 3662.

[44] Feynman, R. P. *Statistical mechanics: a set of lectures*, W.A. Benjamin, Reading, Massachusetts, 1972.

[45] Svidzinskii, A. V. *Spatially inhomogeneous problems in the theory of superconductivity*, Nauka, Moscow, 1982, [in Russian].

[46] Fetter, A. L.; Walecka, J. D. *Quantum Theory of Many-Particle Systems*, McGraw-Hill Book Company, San Francisco, 1971.

[47] Abrikosov, A. A.; Gorkov, L. P.; Dzyaloshinski, I. E. *Methods of quantum field theory in statistical physics*, Dover Publications, New York, 1975.

[48] Engelbrecht, J. R.; Nazarenko, A. *Physica C* 2000, 341, 139.

[49] Szotek, Z.; Györffy, B. L.; Temmerman, W. M. *Phys. Rev. B* 2000, 62, 3997.

[50] Leggett, A. J. *Ann. Phys. (NY)* 1968, 46, 76.

INDEX

A

B

C

D

Q

R

S

T

U

V

W

X

Y